VEHICLE MAINTENANCE AND REPAIR SERIES

ENGINES, ELECTRONICS and RELATED SYSTEMS

Jack Hirst ● Edited by Roy Brooks

THOMSON

LEARNING

...ada ● Mexico ● Singapore ● Spain ● United Kingdom ● United States

THOMSON
LEARNING

Vehicle Maintenance and Repair Series, Level 3 – 3rd Edition
Engines, electronics and related systems

For more information, contact Thomson Learning, Berkshire House, 168–173 High Holborn, London, WC1V 7AA or visit us on the World Wide Web at:
http://www.thomsonlearning.co.uk

British Library Cataloguing-in-Publication Data
A catalogue record for this book is available from the British Library

ISBN 1-86152-805-1

First published 1994 Macmillan Press Ltd
(Vehicle Mechanical and Electronic Systems – Engines, electronics and related systems)
Reprinted 2000 by Thomson Learning
This edition 2002 Thomson Learning

Typeset by Photoprint, Torquay, Devon
Printed in Croatia by Zrinski

Contents

	Preface	iv
	Acknowledgements	v
	Coverage of Standards	vi
	Basic Essentials: Health, Safety and Relationships in the Workplace	vii
1.	Engine – Basic Principles	1
2.	Engine Lubrication	17
3.	Cooling and Heater Systems	32
4.	Spark-Ignition Systems	50
5.	Air Supply and Exhaust Systems	69
6.	Petrol (Carburettor) Fuel Systems	82
7.	Diesel Fuel Systems	99
8.	Battery, Charging and Starter Systems	134
9.	Engine – Mechanical Details, Inspection and Testing	166
10.	Engine Pressure Charging Systems	213
11.	Basic Electronics and Vehicle Electronics	227
12.	Engine Management and Petrol Fuel Systems	251
13.	Air Conditioning	280

Preface

Welcome to this workbook, which along with its companion volume, *Transmission, Chassis and Related Systems* is one of the final books in the Thomson Learning, Vehicle Maintenance and Repair Series. It is great to know that you are probably on the last lap towards gaining Level 3 NVQ/SVQ or similar important qualifications.

Very likely, as with thousands of other students, you have already been helped on your way by using the earlier books in this long established series. Consequently you will know that by properly completing the text and diagrams you will achieve a high standard of essential knowledge. What is more the book will then become a valuable source of reference and provide vital evidence of achievement for your portfolio. One quick point though, don't forget to complete the answers in pencil – just in case of mistakes!

It is worth stressing that although you will now be working at quite an advanced level and often using equipment that is complex and dedicated, it is most important that you understand the basic principles of such items. For this reason the book concentrates where possible on fundamentals rather than the latest fancy variations on what may well be fairly simple designs. This helps you to understand and cope with the many and rapid technological changes that occur in the automobile world. No doubt very wisely you will consult manufacturer's data for any fine detail that might be required.

The author and editor wish you every success with your studies and progression within the motor industry. If you have any comments you wish to pass on to us, please do so via the publisher.

Roy Brooks
Series Editor

Acknowledgements

The editor, author and publishers would like to thank all who helped so generously with information, assistance, illustrations and inspiration. In particular the book's principal illustrator, Harvey Dearden (previously principal lecturer in Motor Vehicle Subjects, Moston College of Further Education); Automotive Training Australia (ATA) for their original chapter on air conditioning; and the persons, firms and organisations listed below. Should there by any omissions, they are completely unintentional.

A-C Delco Division of General Motors Ltd
Autodata Ltd
Automobile Association
Automotive Products plc
Automotive Technology
BBA Group plc
Robert Bosch Ltd
British Standards Institution
Buick Motors
Butterfield Equipment Ltd
Castrol (UK) Ltd
Champion Sparking Plug Co. Ltd
Chevrolet Motors
Chrysler Corporation
Citroën UK Ltd
City & Guilds of London Institute
Comprex
Crypton Technology
Cummins Engine Co. Ltd
Fiat Auto (UK) Ltd
Flute Instruments Ltd
Focal Displays Ltd
Ford Motor Co. Ltd
General Motors Corporation
Gunson Ltd
Leslie Hartridge Ltd
Hepworth & Grandage Ltd
HMSO
Holset Engineering Co. Ltd
Honda UK Ltd
Lancia
Land Rover Ltd
Leyland Bus Ltd

Lotus Group of Companies
Lucas CAV
Lucas Industries plc
MAN–Volkswagen
Mazda
Metalistic Ltd
Mitsubishi Motors
Mobelec Ltd
Mobil Oil Co. Ltd
Motor Industry Training Council (MITC)
Nissan Motors (GB) Ltd
Perkins Group Ltd
Peugeot Motor Co. plc
Renault UK Ltd
Ripaults Ltd
Rover Group Ltd
R S Components Ltd
Saab Great Britain Ltd
Scania Ltd
Seaton Ltd
Sherrat Ltd
Signs & Labels Ltd
Solex Ltd
Sprintex Ltd
Subaru Ltd
Telehoist Ltd
Toyota (GB) Ltd
Unipart Group of Companies
VAG (United Kingdom) Ltd
Vauxhall Motor Co Ltd
Volvo Ltd
Weber Concessionaires Ltd
Zenith Carburettor Co. Ltd

Coverage of Standards

QUICK CHECK UNIT GRID

VEHICLE MAINTENANCE and REPAIR SERIES LEVEL 3

The subject material in chapters covers
Basic Essential Knowledge
for the unit areas indicated.

ENGINES, ELECTRONICS and RELATED SYSTEMS

UNIT NUMBERS and TITLES	Contribute to Good Housekeeping	Ensure Your Own Actions Reduce Risks to Health and Safety	Maintain Positive Working Relationships	Carry Out Routine Vehicle Maintenance	Diagnose Complex System Faults	Rectify Complex System Faults	Enhance Vehicle System Features	Overhaul Units	Identify and Agree Customer Vehicle Needs	Inspect Vehicles
	1	2	3	11	13	14	15	17	18	19
BASIC ESSENTIALS — Health, Safety and Relationships in the Workplace	•	•	•						•	
1. Engine – Basic Principles										
2. Engine Lubrication				•				•		•
3. Cooling and Heating Systems				•	•	•		•		•
4. Spark-Ignition Systems				•	•	•		•		•
5. Air Supply and Exhaust Systems				•				•		•
6. Petrol (Carburettor) Fuel Systems				•	•	•		•		•
7. Diesel Fuel Systems				•	•	•		•		•
8. Battery, Charging and Starter Systems				•	•	•		•		•
9. Engine-Mechanical Details, Inspection and Testing				•	•	•	•	•		•
10. Engine Pressure Charging Systems				•	•	•	•	•		•
11. Basic Electronics and Vehicle Electronics					•					
12. Engine Management and Petrol Fuel Systems				•	•	•		•		•
13. Air Conditioning				•	•	•	•	•		•

For complete syllabus coverage see also the other books in this series – Maintenance and Repair of Road Vehicles Level2 and Transmission, Chassis and Related Systems Level3.

Basic Essentials

Health, Safety and Relationships in the Workplace

Whichever subjects, at whatever level you are studying to obtain NVQ/SVQ qualifications in Motor Vehicle Work, you must be successful in the first three units:

1 Contribute to Good Housekeeping

2 Ensure Your Own Actions Reduce Risks to Health and Safety

3 Maintain Positive Working Relationships

Even if you are already familiar with this area of work, the pages in this section revise and reinforce these vital units. Their contents consist of items/actions which you should observe and carry out every day of your working career.

Good Housekeeping – keeping the workshop clean, tidy, and safe   viii
Workshop Resources – looking after equipment, power, and time   x
Health and Safety at Work – what to be aware of   xi
Accidents and First Aid – what to do if something happens   xi
Hazards – what to watch for   xii
Personal Protection – clothes and equipment   xii
Safe Handling – of loads, of equipment, and of harmful substances   xiv

Health and Safety Signs – what they mean   xvi
Emergency Procedures – fire alarms and fire extinguishers   xvi
Positive Working Relationships – how to develop good relationships with colleagues   xviii
Working as a Team – knowing your own job and what other people do   xviii
Organisational Structure – knowing who does what in the company   xix
Communication – spoken, written, telephone, and non-verbal   xx

GOOD HOUSEKEEPING

Maintaining a clean work area

We are all impressed when we see a clean and shiny car, even though we know it would work just as well dirty. In the same way customers will be impressed if you keep your workshop clean and tidy. No one wants to see a dirty workshop, a cluttered parts department, an untidy forecourt, or a patch of oil on the floor.

Remember – there are benefits in keeping the workshop clean:

- Passing customers who need work done may be attracted in.
- Regular customers will be happy to keep coming back.
- Staff will work better.
- Accidents are less likely to happen.
- Work will be completed faster.
- You are less likely to lose tools and parts.

Housekeeping routines

Each workshop will have a system to keep the place clean and tidy. As an employee you may have to do some simple cleaning or tidying up as part of your duties.

A large company may employ cleaners; in a small one the work will be shared between the staff. All technicians are likely to be responsible for cleaning up after their work. Some jobs are particularly dirty – for example, jobs on exhausts or suspensions always leave dust and dirt on the floor.

Make sure you know your firm's housekeeping routines. Here is a typical daily housekeeping routine for a small garage workshop.

OPENING

- Move parked vehicles away from the work areas.
- Check that the lifts and floor area are clean and free from obstructions.
- Check that all tools and equipment are clean and tidy.
- Check the reception counter is tidy, and that the sales computer equipment is working.
- Check that the reception area and seating are straight and tidy.
- Check the special workshop tools and equipment such as the air line, the tyre-remover and the wheel balancer are working.

DURING THE DAY

- Keep the lift and the floor clean.
- After use, put special tools back where they are normally kept.
- Make sure that items such as wheels and removed tyres do not obstruct work areas or pathways.
- Make sure that the reception area is kept tidy, with ashtrays empty and magazines straight.

CLOSING

- Put away neatly all special tools.
- Lock up your personal toolbox.
- Check that any customers' vehicles are secure.
- Clean and tidy the work area.
- Switch off power to equipment.
- Tidy reception and empty the till.

Delivery of goods

When goods are delivered, do not place or leave them where they would block walkways or exits.

Cleaning

Cleaning equipment should be kept in a separate store, as many chemicals are highly concentrated.

Always read the instructions on the labels before using them (see COSHH regulations). If specialised cleaning is required, your employer will provide protective clothing.

When you have finished cleaning, put the cleaning equipment and unused chemicals back in the store.

HEALTH & SAFETY

If you need to use a hazardous cleaning material, read the label on the container. This will tell you how to use it safely, and what to do if you *do* have an accident – for example, if the cleaning material touches your eyes or skin.

OXIDIZING

TOXIC

IRRITANT

CORROSIVE

Make a list of the cleaning materials available in your garage. Think about:

- *Personal cleanliness* – e.g. barrier cream, hand-cleaning material.
- *Equipment* – e.g. lifts, trolley jacks, wheel-balancing machines, special and personal tools.
- *The workshop* – e.g. the floor area, car bays, walls, windows, lighting.

Emergency cleaning

Sometimes you will need to clean up after a breakage or spillage. Some oil might be spilt, for example, or some glass broken on the floor.

Breakages and spillages must be cleaned up immediately. If this is not done, someone may be injured. Also, the firm could be in breach of the Health and Safety at Work Act (see page xi). If an accident happens, the firm may be fined.

Disposing of dangerous waste material

All workshops produce dangerous waste materials – dirty engine oil and filters, scrap exhausts and batteries, broken or scrap plastic or metal components, and waste paper.

Items must be disposed of in different ways. Usually this is decided by the local council, who pass by-laws. Refuse disposal requirements differ from place to place.

Some types of dangerous material must be kept separate. They will be collected by specialist agencies, or taken to the local refuse collection point.

HEALTH & SAFETY

While cleaning, place cones and notices to warn others. Section off areas that could be dangerous, such as slippery floors.

⚠️

**DANGER
Cleaning in
progress**

All must take note of the Environmental Protection Act.

WORKSHOP RESOURCES

Every workshop has many resources. You need to be aware of what they are, and how you can make best use of them. Resources include:

- *Special workshop tools and equipment fixtures*, such as vehicle lifts, the air compressor, the wheel-balancer, steering alignment tools, the headlamp aligner, and so on.
- *Stock* in the stores – different types of tyres, exhausts, shock absorbers, batteries, oil, and so on.
- *Fixtures and fittings* in the reception area, the staff dining area, and elsewhere.
- *Utilities*, such as electricity, gas, water and the telephone.
- *Space* available – to work, to store parts, to park and display vehicles.
- *Time* – often the firm's most valuable resource.

Use resources to the best advantage

TOOLS

Use tools and equipment safely and properly. Avoid damage, and don't risk your own health and safety, or anyone else's.

UTILITIES

Electricity, gas and telephone calls are costly. Waste will reduce the firm's profit. To save energy, your firm will probably have some automatic timing controls fitted.

CONSUMABLE ITEMS

Do not waste consumable items, even if they are small. After fitting an exhaust system, for instance, return to stock any unused components such as nuts, brackets or rubber mounting rings.

Security

Make sure that parts are kept as safe as possible. If visitors wander around the workshop and are not observed, they could steal things. Theft by staff may also occur.

Taking goods, equipment or money without permission is always theft. Theft by staff is gross misconduct, and can lead to dismissal.

Do not leave keys in unattended cars. It is not uncommon to have cars driven away from the forecourt.

If you see someone acting suspiciously, ask them what they are doing. If the answer is not satisfactory, tell a senior member of staff immediately. Do not put yourself at risk.

Using resources economically

AVOID WASTING POWER

- Turn off lights when they are not needed.
- Keep workshop doors shut, to keep heat in.
- Report faulty components – for example, a leaking air line that causes the compressor to keep switching on.
- Turn off water when you are not using it, especially when washing cars.

USE SPACE SENSIBLY

Workshop space is expensive. Costs include rates, taxes, heating and lighting. Do not waste this space!

- Place vehicles so that you can work properly, but so that they take up minimum space.
- Keep the working space around ramps and in gangways clear of obstructions.
- Clear away quickly when a job is finished.

TIME

You rightly expect to be paid for your time at work. So you must play your part and help the firm to make a profit.

- As a trade trainee you are unlikely to work as quickly as a skilled technician. But aim to build up your speed.
- Work steadily, but not slowly. Time wasted is almost impossible to make up. You may have a bonus at stake!
- Organise yourself. Gather tools, information and equipment before you start a job.
- Complete jobs properly. Putting things right later is costly.

HEALTH AND SAFETY AT WORK

You must always take care of your own health, hygiene and safety. Here we look at health and safety matters that affect you at work and explain some of the regulations. These exist to make sure that you work in good, safe conditions. All major regulations have developed from the legislation in the **Health and Safety at Work Act 1974**.

Regulations

In the United Kingdom and Europe, health and safety matters are covered by the **Management of Health and Safety at Work Regulations 1992**.

As you train you will need to learn and understand the safety regulations that apply to your job. You are as responsible as your employer for following workshop safety regulations. You will find copies of the regulations displayed in the workshop. For your own safety, **read them**.

Duties

Under the Health and Safety at Work Regulations, **you** have certain duties. Bear these in mind as you work.

- Take reasonable care of your own health and safety.
- Take reasonable care for the safety of other people who may be affected by your actions.
- Work with your employer to keep safety rules.
- Report any accidents, hazards, or damage to equipment.

Warning

If you do not follow the Health and Safety at Work Regulations, you could be taken to court. For example, suppose you were welding an exhaust pipe. If you did not wear the goggles provided, you would be breaking the law, because you would not be taking reasonable care of your own health and safety.

ACCIDENTS AND FIRST AID

An accident involves something that is unexpected and unplanned. One or more people may be injured.

If the accident is minor, it may just be inconvenient. If it is serious, though, it could affect you for the rest of your life.

Accidents may be caused if you:

- Do not know the dangers involved in what you are doing.
- Daydream.
- Do not take enough precautions.
- Fool around.

Accidents may also occur because of faulty equipment or bad work conditions:

- Unsafe tools.
- Unguarded machinery.
- Poor ventilation.
- Poor lighting.

In your place of work, someone will have been specially trained as a first-aider. Normally this person will give first-aid. If the accident is minor, you may be able to help.

Whenever an accident happens, it is important to record the details in a special accident book.

Every first-aid kit should contain a card giving advice on what to do if anyone is injured.

HAZARDS

A hazard is anything that might cause an accident or injury.

Look around your workshop. You will probably see at least one hazard, perhaps more. Some can quickly be removed. Examples are:

- Wheels left lying on the floor, after being taken off a vehicle.
- Oil spilt on the floor.
- A trolley jack handle left lying where someone could trip over it.

Some hazards are always present. Warning notices or guards, or both, keep the risk to a minimum. Examples are:

- Brake tester rollers.
- A wheel balancer.
- A stand drill.

PERSONAL PROTECTION

Various equipment is available which can be worn or held by people at work, protecting them from risks to health and safety. This equipment is covered by the **Personal Protection Equipment at Work Regulations 1992**. All Personal Protective Equipment (PPE) in use at work should carry the CE mark and where appropriate should comply with a European Norm (EN) standard. The Regulations don't include ordinary working clothes that do not specifically protect the health and safety of the wearer.

Personal presentation

Take care in what you wear.

To protect yourself and your clothes, it is sensible to wear:

- One-piece overalls (a boiler suit).
- Stout footwear (preferably with steel toe-caps).
- A suitable cap or bump cap.

Do **not** wear:

- Loose or torn overalls (especially if the sleeves are loose or torn).
- Rings or watches.
- Trainers or similar.
- Long hair (unless protected by suitable headgear).

HEALTH & SAFETY

Hazard
"a hazard is something with potential to cause harm"

Risk
"a risk is the likelihood of the hazard's potential being realised"

HEALTH & SAFETY

Your personal presentation at work should:
- help to ensure the health and safety of yourself and others
- meet legal requirements
- be in accordance with workplace policies

Specialised personal protection equipment

In your work you will need specialised equipment to protect you. Examples include dark-tinted glass goggles when welding, masks when painting, waterproof clothing when steam cleaning, or high-visibility clothing when going out to vehicle breakdowns.

As you do vehicle repair and maintenance work you will sometimes need special equipment to protect the top of your head, your eyes, your ears, your hands and feet, and your breathing.

SAFETY CAPS

Bump caps protect your head from banging on the underside of a vehicle when you work under a ramp.

Soft caps keep your head and hair clean. They also prevent long hair from catching in revolving parts, such as drills on the bench, or engine drive belts under the bonnet.

Bump cap

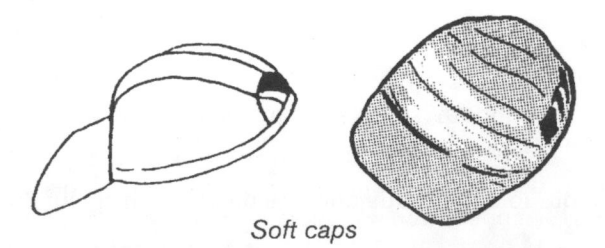

Soft caps

EYE PROTECTORS

An accident to the eyes can be very painful and may result in blindness.

Spectacles protect you from rust or dirt falling off the car. Some have side shields and can be adjusted to fit your face.

Goggles protect you from dust and chemicals. They are used, for example, when sanding body filler off bumped car wings.

Welding goggles protect your eyes from the bright glare of the welding flame.

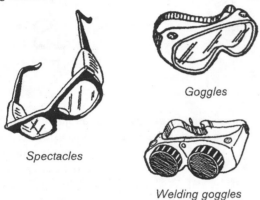

Goggles

Spectacles

Welding goggles

EAR PROTECTORS

Ear muffs protect your ears from damage when there is a loud, continuous noise.

Ear plugs are as effective as, and in some cases more effective than, ear muffs!

Earplugs

Ear muffs

MASKS

Face masks protect your lungs from dust (some of the dust may be toxic.) They use special moulded pads made from cotton gauze or special filter paper.

Gas respirators are used in vehicle paint shops. The paint fumes may be toxic.

Face mask

HAND PROTECTION

Industrial gloves should be used when moving rough or heavy parts. They protect your hands and wrists from cuts, scratches and burns.

Heat-resistant gloves should be worn when working on items such as a hot exhaust or radiator.

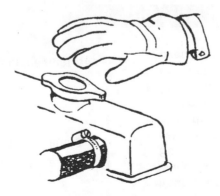

Heat-resistant gloves

SKIN PROTECTION

Before you start work, put barrier cream on your hands. If you have sensitive skin you might easily develop an infection such as dermatitis. Thin plastic gloves (like surgical gloves) can be worn to prevent contact with fuel and oil.

After work, clean your hands with an antiseptic hand cleaner. Rub this on your hands before you get them wet.

Keep your overalls clean, and do not put dirty rags in your pocket. Oil might pass through your clothes and onto your body, and you might develop a skin infection.

FOOT PROTECTION

Safety boots protect your feet and toes from falling objects. In a workshop there is also a risk that a car might run over your feet!

TOTAL PROTECTION

Total waterproof protection will sometimes be needed.

- When working at a car wash or valeting firm, wear waterproof clothing.
- When working on breakdowns, wear high-visibility clothing.
- When paint-spraying a car, wear Tyvek overalls. These have elasticated hoods, cuffs and ankles. Wear a gas respirator, too.

SAFE HANDLING

Moving loads

A load is any heavy object that must be moved, whether by hand or by lifting equipment.

CORRECT HANDLING TECHNIQUES

When you have to lift something big or heavy, you need to lift it in the right way. Look at the drawings and read the numbered instructions.

LIFTING A HEAVY PART FROM THE FLOOR

1 Stand as close to the load as possible. Spread your feet.
2 Bend your knees and keep your back in a straight line. Do not bend your knees fully, as this would leave you with little lifting power.
3 Grip the load firmly.
4 Raise your head.
5 Lift by straightening your legs. Keep the action smooth.
6 Hold the load close to the centre of your body.

UNLOADING ONTO A BENCH

1. Bend your knees to lower the load. Keep your back straight, and weight close to your body.
2. Be careful with your fingers as you set the load down.
3. *Slide* the load into position on the bench. Push with your body.
4. Make sure that the load is secure, and that it won't tip, fall or roll over.

HEALTH & SAFETY

Avoid personal injury. Do not lift anything too heavy for you – about 20 kg is a recommended amount.

If using lifting gear, never exceed the Safe Working Load (SWL).

✓ Safe working load ▢ kg

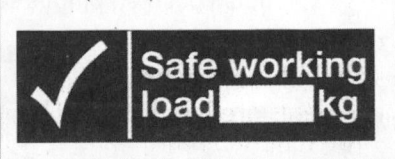

Handling harmful substances

Workshops often store dangerous chemicals. Some could catch fire; some could even explode when handled. Others are corrosive or caustic, and could damage your skin.

There are regulations about the **Control of Substances Hazardous to Health (COSHH)**. These state that every hazardous substance must be described on a health and safety data sheet. The sheet gives details of safe handling, and says whether protective equipment should be worn.

Safe use of garage machinery and equipment

RAMPS AND JACKS

In the workshop you will often use vehicle lifts and trolley jacks. Make sure you know how to use them safely.

The diagram opposite shows basic precautions you should take when working under a lift. Here are three extra precautions:

- Do not exceed the lift's safe working load (SWL).
- Before raising a car, check that the radio aerial, bonnet and boot lid are down. They could hit lights, beams or the roof.
- Before lowering, make sure that all tools and old parts have been removed.

COMBUSTIBLE MATERIALS

Some liquids or chemicals found in a garage catch fire very easily. Petrol is the most obvious example. Vapour from such chemicals could be ignited by a spark – even the tiny spark when a light switch is operated.

Such fluids must be stored in fireproof containers. These are designed to prevent leakage and evaporation.

ELECTRICAL SAFETY

The main dangers caused by electricity are:

- Fire – due to cables being overloaded, overheating, or loose connections.
- Electric shock – due to touching a live circuit.

Unguarded cables or connections, like those in the diagram opposite, could cause a fire or a shock.

In industrial premises, all electrical equipment must be checked regularly by a qualified person.

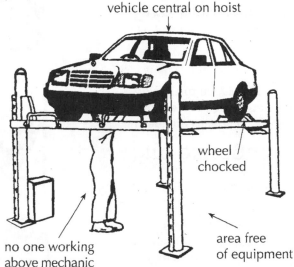

vehicle central on hoist

wheel chocked

no one working above mechanic

area free of equipment

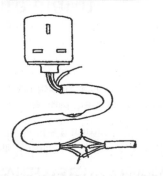

HEALTH & SAFETY

For your own safety, make sure the equipment you use has been checked and is safe.

HAND TOOLS

Hand tools are spanners, sockets, screwdrivers, pliers, hammers, chisels and files. To work safely with them use your common sense, know which tool to use, and follow safe procedures.

In a workshop the most common small injuries are cut fingers or skinned knuckles and fingers. Usually these are due to the misuse of a hand tool.

ROTATING MACHINERY

All high-speed rotating machinery, such as wheel balancers, drills and grindstones, should have guards fitted.

COMPRESSED-AIR EQUIPMENT

Compressed air is dangerous if misused. Before you use flexible pipe extensions, make sure that the quick-release couplings are fully engaged.

When working with compressed air:

- Never direct it onto any part of your body.
- Never use it to blow away brake dust (or any other type of dust).
- Never use it to clear dirt or filings off benches.
- Never use it to clean ball and roller bearings (by spinning them).

HEALTH AND SAFETY SIGNS

All public places, including workshops, must display safety signs to warn people of dangers. If you look around buildings you will see such signs. Some will be so familiar that you hardly notice them.

By law (BS 5378 and BS 5449) safety signs must clearly show what they mean. There are different shapes, colours, and symbols or words.

There are four types of safety signs, and fire signs.

Prohibition

A red circular band and a cross bar.

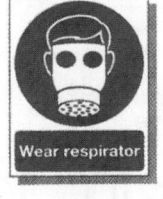

Mandatory

A blue circle with a symbol inside.

Warning

A yellow triangle with a thick black border.

Safe condition

A green square or rectangle with a symbol inside.

EMERGENCY PROCEDURES

Emergencies may be caused by many things: a workshop accident, a fire, a spillage of a flammable or hazardous substance – even a bomb scare.

If you are sure there *is* an emergency, sound the alarm, evacuate the building and ring **999** for the emergency services. (See next page.)

Fire alarm

In a small workshop, the fire alarm may be simply shouting '*Fire*'. In a large workshop an automatic alarm may be linked to the fire station, for immediate action.

If a workshop employs more than five people, it must have an emergency evacuation procedure. It must also have a building plan, and signs that show where to find:

- Fire extinguishers.
- Fire exits.
- Assembly points.
- First-aid points.

Fire extinguishers

There are several types of fire extinguisher, suitable for different kinds of fire. Your garage should have extinguishers to fight fuel fires and electrical fires, as well as ordinary water extinguishers.

Fire equipment

A red square or rectangle with a symbol or text.

Extinguisher contents

The different kinds of extinguisher have different contents.

- **Water extinguisher** Water will kill the heat and put out the fire. This should be used for wood and paper fires. Do not use it for electrical fires: you could get an electric shock. Do not use it for petrol or oil fires; burning fuel will float on the water.
- **Foam extinguisher** The foam is water-based. It smothers the fire which goes out because there is no oxygen. This can be used with flammable liquids. Do not use this in a garage.
- **CO₂ extinguisher** This produces carbon dioxide gas, which removes the oxygen around the fire. However, because CO_2 does not remove the heat, wood and paper could re-ignite later.
- **Dry powder extinguisher** The powder is a fire-retardant dust. This covers the fire like a blanket.
- **BCF extinguisher** This smothers the fire with a blanket of heavy vapour. It is very clean, and leaves no deposit.

New colours

Since January 1997 the British Standard for fire extinguishers has been BS EN 3. Under this standard, all fire extinguishers must be coloured *red*. However, 5% of the surface area may be colour-coded using the colours many people are already familiar with. The table shows types of fires and the extinguishers recommended for fighting them.

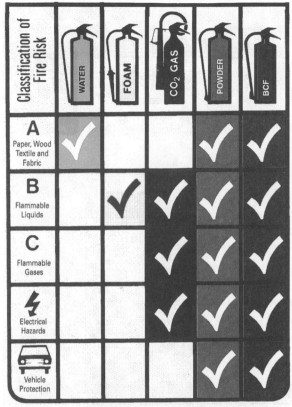

Classification of Fire Risk	WATER	FOAM	CO₂ GAS	POWDER	BCF
A — Paper, Wood Textile and Fabric	✓			✓	✓
B — Flammable Liquids		✓	✓	✓	✓
C — Flammable Gases			✓	✓	✓
Electrical Hazards			✓	✓	✓
Vehicle Protection				✓	✓

HEALTH & SAFETY

A fire blanket can be wrapped around someone who is burning. It smothers the fire.

Fire blanket

Emergencies

In an emergency, for **fire, police** or **accident**, press or dial **999**.

1. Tell the operator which service you want.
2. Wait for the operator to connect you.
3. Tell the emergency service:
   a. Where the trouble is
   b. What the trouble is
   c. Where you are
   d. The number of the phone you are using.
4. Let the person at the other end of the telephone ask the questions.
5. Don't hang up until the service has all the information it needs.

WARNING

Never make a false emergency call. It is against the law, and you could risk the lives of others who *really* need help.

You can be traced immediately to the telephone where the call came from.

POSITIVE WORKING RELATIONSHIPS

Working with others

There are several different kinds of relationships. You have a family relationship with a relative – a parent, a child, an aunt or a cousin. With people you like at school or college, you have a friendly relationship. You may have a romantic relationship with a girlfriend or boyfriend.

Working relationships develop with the people who work alongside you. As you interact in the workshop with the manager, other mechanics and fitters, or your supervisor, you build working relationships with them.

Good working relationships are very important to the success of every business. You may not be friends with all your colleagues; occasionally you may even dislike some of them. But to help the business run smoothly, you must get on with all of them professionally.

WORKING AS A TEAM

To make the company successful, all of its employees must work together. They must co-operate, like members of a football team: this is teamwork.

If the firm does well, and employees get on with each other and trust each other, there will be a good feeling in the workplace, and people will be enthusiastic about their jobs. This feeling is called good morale.

When everyone works hard, and no one wastes time or resources, the firm will be efficient. By being efficient, employees will get a lot done – they will be productive.

If a team with good morale works efficiently and productively, customers will be satisfied and pleased to come again. They are also likely to recommend the firm to others, so it will gain a good reputation. And this in turn will bring more business, and the firm will become even more successful, and will grow. It will gain a good company image.

The company's aim

It is important to be aware of the aim of the company for which you work. This may be set out in what is known as a mission statement. This explains the main objectives of the company: the reasons why it is trading, and what it hopes to achieve.

All firms are in business to make a profit. This is shared between the owners or any shareholders. A profitable business is likely to be a successful one, and a successful business can offer its workforce job security.

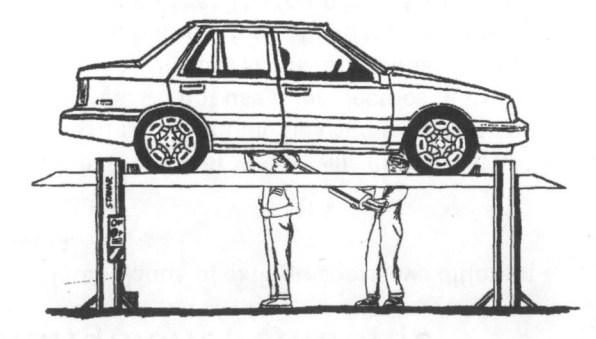

ORGANISATIONAL STRUCTURE

The organisational structure describes who does what in the company. It describes each person's duties and level of authority. It also shows everyone's working relationships, from the managing director to the most junior employee.

Job description

A job description names a job title and states the duties and responsibilities of a particular job.

Each employee should be aware of their own role and the roles of others. Disputes may arise if anyone is uncertain about who does what.

An organisation chart such as the one below shows levels of authority and responsibilities. Vertical links show the line of authority from senior staff downwards. Horizontal links indicate people with equal status and authority in the company.

Building good relationships

Here are some ways in which you can build good working relationships with your colleagues.

- Recognise that there are differences in personality and temperament.
- Treat colleagues politely and with respect.
- Co-operate, and assist willingly with requests.
- Talk with colleagues about problems, changes or proposals.
- If something you are doing goes wrong, or if you break something, tell your supervisor straight away.
- When a working relationship breaks down, be honest and fair, and try to put things right.

Here are a few examples of the kind of things that can upset good working relationships

- In everyday conversation you may discuss your social life, sports, films and so on. You may find that a workmate has opinions on some topics which are different from yours. Never allow these differences to spoil your working relationship.
- Sometimes you may find colleagues who lack interest or enthusiasm, who are lazy or incompetent, who keep being absent, and so on. If people do not 'pull their weight' in the workplace, this can cause anger and frustration among other members of the team.
- Beware of anyone who ignores company rules and regulations, or safe working practices. This behaviour can create problems or even dangers for everyone else in the company.
- Personal appearance and hygiene are important. Employees who do not bother with these may upset colleagues, and this may affect the performance of a workforce.
- Managers and supervisors should not show favouritism and should pay everyone fairly.

The company image depends in part on the way its workers are judged by customers. Customers will see your work, and also how you behave and interact with your colleagues.

COMMUNICATION

When problems and misunderstandings do arise, it is often because of poor communication. Good communication is an essential part of every working day, for employees at all levels.

In a typical garage, there will be daily communication between colleagues within a department, between departments, between management, supervisors and shop-floor staff, and between customers and reception staff. The company will also communicate with suppliers, subcontractors, vehicle manufacturers, advertising agencies, banks, accountants, lawyers, the council, and so on.

Trust and goodwill with customers will be very much helped if you practise good, clear communication. Misunderstandings can easily make customers go elsewhere. In Britain, customers who are not satisfied tend not to complain – they just don't come back. So you may never find out that something is wrong – until, perhaps, there is no work!

Non-verbal communication

It is not only your words that communicate – so too do your facial expression and the way you stand and move. Customers will notice your body language: for example whether you smile and look directly at them or slouch and avoid their eyes.

Problems in communication

Some things can cause problems in communication.

LACK OF COMMUNICATION

Problems can arise if people are not told things they need to know. Changes to systems, safety issues, shop-floor problems and the like must be passed on to the right people without delay.

Delays in getting information to staff may be caused by pressure of work or different hours of work (such as part-time, full-time, or shift work).

INCORRECT COMMUNICATION

Problems can also arise if information given is wrong. A good example of this is when a spoken message is passed on and wrongly remembered – always write down messages for people, while the information is clear in your mind. Similarly, if a job sheet lacks some detail this causes confusion.

If a message is passed orally via several people, it may become totally changed!

Methods of communication

This section is particularly concerned with the effects of communication on working relationships. The method of communication chosen will depend on the situation.

1 *Direct discussion* – face-to-face conversation.
   a With a customer, to find out their requirements and to advise them.
   b Between colleagues, about a particular job.
   c With management and supervisory staff, to talk over procedures and problems.
2 *Writing*
   a Recording customer and vehicle details.
   b Preparation of job sheets. Inter-department memos.
   c Letters. Reports. Manuals. Notices.
3 *Telephone*
   a Quick communication with customers, suppliers and others.
   b Between departments and colleagues.
   c Quick contact with emergency services.

Chapter 1

Engine – Basic Principles

Main engine components	2	Combustion processes	11
Four-stroke cycles	3	Cylinder head – combustion chamber shapes	
Two-stroke cycle – spark-ignition	4	(spark-ignition)	11
Two-stroke cycle – diesel (compression-ignition) engine	5	Combustion sequences (spark-ignition)	12
Engine terminology	6	Combustion chambers used in diesel (CI) engines	13
Cylinder swept volume	7	Combustion in a diesel (CI) engine	14
Compression ratio	7	Valve timing	15
Crankshaft-to-camshaft movement ratio	8	Diesel (CI) four-stroke cycle – valve timing	16
Use of more than one cylinder	8	Valve-port timing diagrams two-stroke spark-ignition	16
Cylinder arrangements, firing orders	9	Two-stroke compression-ignition	16

MAIN ENGINE COMPONENTS

Examine the drawing of the
standard push-rod, four-
cylinder engine shown, and
name the parts shown.

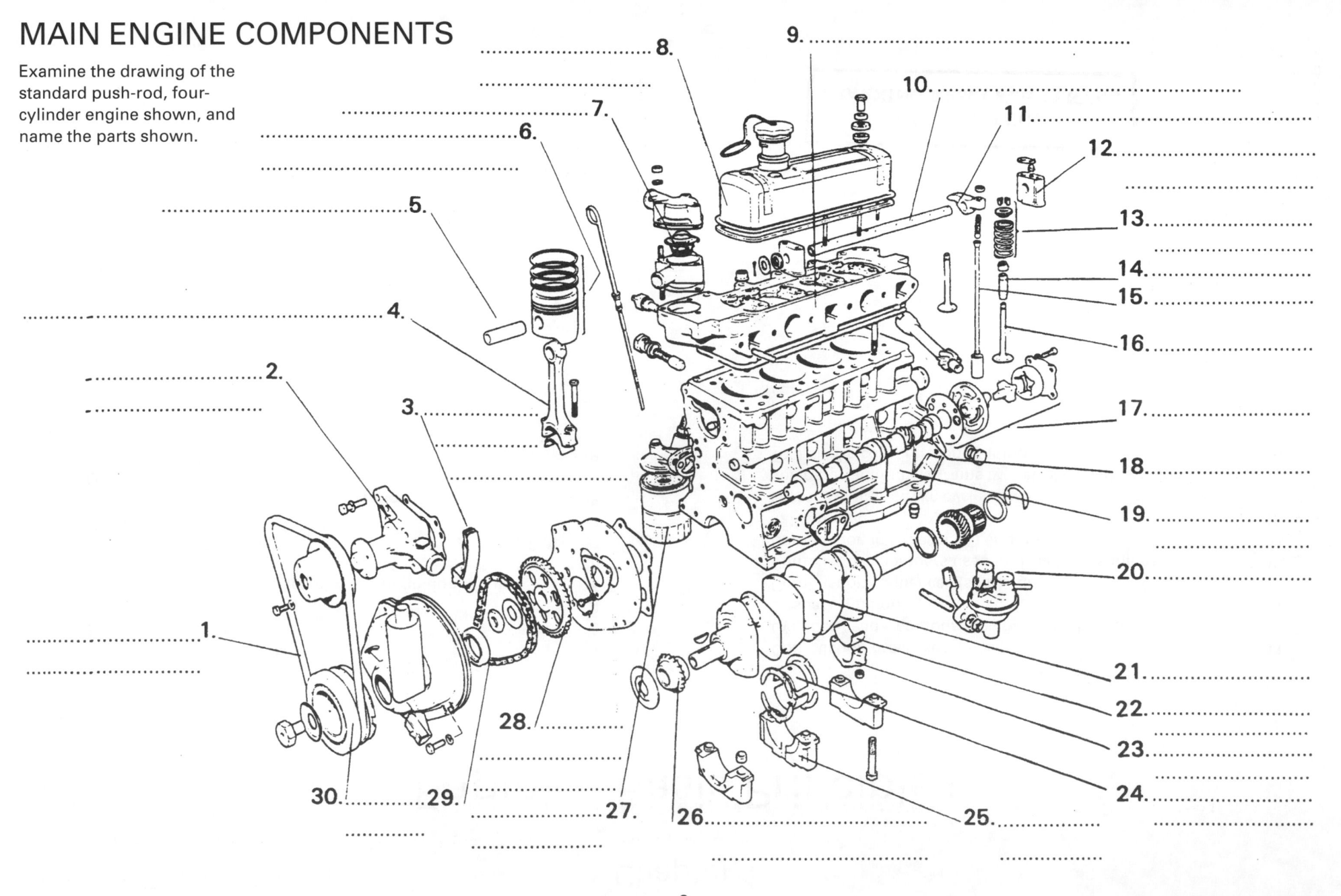

FOUR-STROKE CYCLES

The basic function of an engine is to convert chemical energy to mechanical energy and to produce usable power and torque.

Describe, emphasising their basic differences, the four-stroke petrol (spark-ignition) and diesel (compression-ignition) engine operating cycles. Mention typical timings and compression ratios. Indicate the gas flow into and out of the cylinders.

Petrol (**spark-ignition**)

Induction

....
....
....
....

Compression

....
....
....
....

Power

....
....
....
....
....

Exhaust

....
....
....

Diesel (**compression-ignition**)

Induction

....
....
....
....

Compression

....
....
....
....

Power

....
....
....
....

Exhaust

....
....
....

TWO-STROKE CYCLE – SPARK-IGNITION (crankcase compression type)

By making use of both sides of the piston, the four phases – induction, compression, power, exhaust – are completed in two strokes of the piston or one crankshaft revolution.

No valves are used, the piston itself acts as a valve covering and uncovering ports in the cylinder wall.

Describe the Two-Stroke operational cycle.

FIRST STROKE – Piston moving down.

(a) Events above the piston:

....

....

....

....

....

(b) Events below the piston:

....

....

....

SECOND STROKE – Piston moving up.

(a) Events above the piston:

....

....

(b) Events below the piston:

....

....

....

Name the main parts and show the direction of the crankshaft rotation:

(a) Show fuel entering crankcase.

(b) Show fuel transferring to cylinder and exhausting.

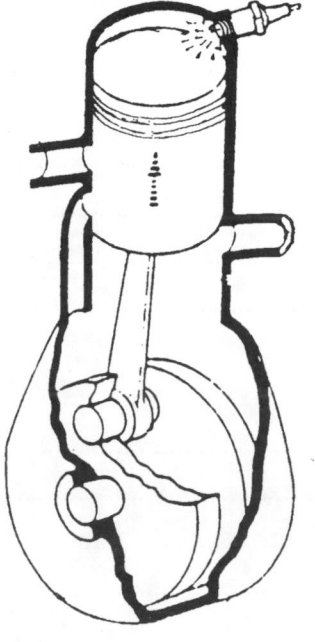
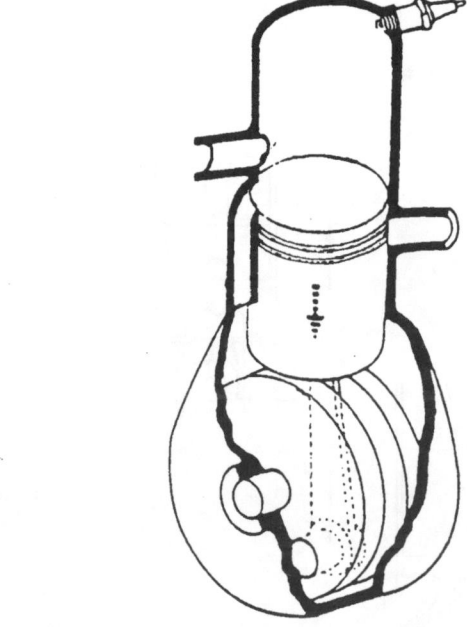

(c) Complete the summary table shown below:

Summary of two-stroke cycle		
Stroke	**Upward**	**Downward**
Events above piston	....	....
Events below piston	....	....

TWO-STROKE CYCLE – DIESEL (COMPRESSION-IGNITION) ENGINE

Describe, with the aid of the diagrams provided, the cycle of operations during the two strokes.

Indicate air/gas flow on each diagram.

First stroke – piston moving up

....

....

....

....

....

....

....

....

....

....

....

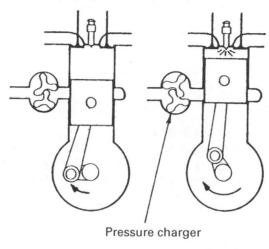

Pressure charger

Second stroke – piston moving down

....

....

....

....

....

....

....

....

....

....

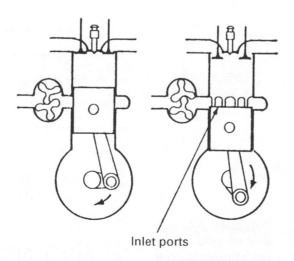

Inlet ports

Why is it necessary to pressure-charge this type of engine?

....

....

....

....

....

....

....

....

....

....

....

....

....

List three advantages gained by using this type of pressure-charged engine.

....

....

....

....

....

....

....

5

ENGINE TERMINOLOGY

Explain the meaning of the following terms:

tdc ....

....

bdc ....

....

Bore ....

Stroke ....

....

Cylinder capacity ....

....

Engine capacity ....

....

Complete the lettered drawings below to show:
on (a) cylinder bore; on (b) tdc and (by shading) the clearance volume; and on (c) the stroke and (by shading) the swept volume.

(a)   **(b)**   **(c)**

The swept volume of a cylinder in a four-cylinder engine is 249 cm³.
Calculate the total volume of the engine.

....

....

....

....

This engine would be known as

.... litre engine.

The swept volume of a cylinder in a six-cylinder engine is 332 cm³.
Calculate the total volume of the engine.

....

....

....

....

This engine would be known as a

.... litre engine.

The total volume of a four-cylinder engine is 1498 cm³.
Calculate the swept volume of one cylinder.

....

....

....

....

This engine would be known as a

.... litre engine.

The cross-sectional area of the piston crown is 48.5 cm² and the stroke is 12 cm.
Calculate the swept volume of the cylinder and the capacity when it has six cylinders.

....

....

....

This engine would be known as a

.... litre engine.

CYLINDER SWEPT VOLUME

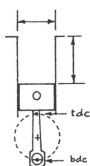

The volume of an engine cylinder is found by multiplying the area of the cylinder by the distance moved by the piston (stroke).

Area of cylinder $= \pi r^2$

where $r = $ cylinder bore/2

Swept volume $= $ ....

Insert the appropriate dimension abbreviations on the drawing.

Example. Calculate the swept volume of an engine cylinder having a bore diameter of 84 mm and a stroke of 90 mm.

Take π as $\dfrac{22}{7}$

(Since engine capacity is quoted in cubic centimetres (cm³), sometimes written as ccs, the basic dimensions should be first converted to cm.)

Therefore bore 84 mm = 8.4 cm
stroke 90 mm = 9.0 cm

Swept volume $= \pi r^2 \times$ stroke

$= \dfrac{22}{7} \times 4.2 \times 4.2 \times 9$

$= \dfrac{22}{1} \times 0.6 \times 4.2 \times 9$

$= 498.96 \text{ cm}^3$

If this cylinder was from a four-cylinder engine it would be a
.... litre engine.

1. Calculate the swept volume of an engine cylinder having a bore diameter of 70 mm and a stroke of 100 mm.

2. Calculate the swept volume of an engine cylinder having a bore diameter of 80 mm and a stroke of 70 mm.

3. Calculate the capacity of a four-cylinder engine whose bore and stroke are both 90 mm.

COMPRESSION RATIO

The engine cylinders below have a compression ratio of ....

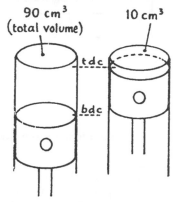

The clearance volume is ....

....

....

....

The compression ratio is the proportion by which ....

....

....

....

This may be expressed as

Compression ratio $= \dfrac{\text{Total volume}}{\text{Clearance volume}}$

$= \dfrac{\text{Swept volume + Clearance volume}}{\text{Clearance volume}}$

$$CR = \dfrac{SV + CV}{CV}$$

1. Calculate the compresson ratio of a cylinder when the swept volume is 720 cm³ and the clearance volume is 90 cm³.

....

....

....

....

2. Calculate the compresson ratio of a cylinder when the swept volume is 518 cm³ and the clearance volume is 74 cm³.

....

....

....

....

Investigation

Examine data books to determine the compression ratio of various engines.

Vehicle model	Engine capacity	CR

CRANKSHAFT-TO-CAMSHAFT MOVEMENT RATIO

The crankshaft-to-camshaft movement ratio in all four-stroke engines is................

On two-stroke CI engines using exhaust valves the movement ratio is..................

On simple camshaft drive arrangements this ratio can be proved mathematically by either counting the number of teeth on each gear or by measuring the gears' diameters and dividing the **driven** gear value by the **driver** gear value.

Calculate the movement ratios of the engines below by:

1. Counting the number of teeth on each gear.

2. Measuring the diameter of each gear wheel.

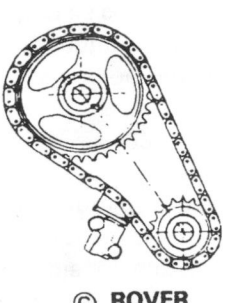

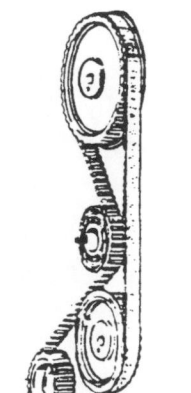

© ROVER

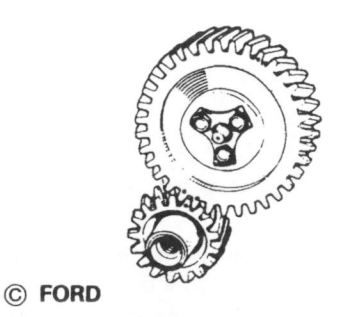

© FORD

Examine engines with drive arrangements similar to those shown above and complete the table.

| Engines examined | Number of teeth | | Gear ratio |
	Camshaft gear	Crankshaft gear	

USE OF MORE THAN ONE CYLINDER

Two of the main reasons why a conventional vehicle uses an engine with more than one cylinder are:

1. A multi-cylinder engine has a higher power-to-weight ratio than a single-cylinder engine.

2. With multi-cylinder engines there are more power strokes for the same number of engine revolutions. This gives fewer fluctuations in torque and smoother power output.

State three other possible reasons for using multi-cylinder engines.

....

....

....

....

....

....

....

....

....

Complete the table below:

Number of cylinders	Engine speed rev/min	Total number of firing strokes/min (four-stroke cycle)
1	1000	
2	1500	
3	2000	
4	3000	
5	2000	
6	3000	

CYLINDER ARRANGEMENTS – FIRING ORDERS

The most common way of arranging the position of cylinders for multi-cylinder engines is in-line.

Name vehicles which use in-line engines:

Vehicle	No. of cylinders	Engine capacity	Firing order
	2		
	4		
	5		
	6		

TWIN-CYLINDER

The cranks may be arranged in two ways.

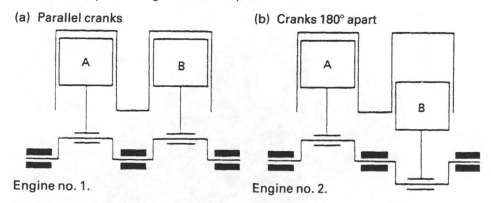

(a) Parallel cranks   (b) Cranks 180° apart

Engine no. 1.   Engine no. 2.

Referring to the above engines	Eng. 1	Eng. 2
What stroke will piston B be on when A is on the power stroke?		
How many crankshaft degrees will the intervals be between the power impulses?		

FOUR-CYLINDER

Complete the line diagram to show a four-cylinder in-line engine:

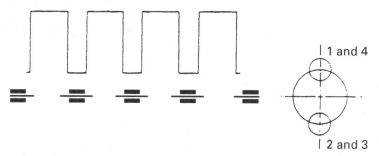

1 and 4

2 and 3

Considering a four-cylinder four-stroke in-line engine, if the firing order is 1342 what stroke would the following be on?

(a) When no. 2 cyl. is on the power stroke, no. 4 cyl. is on ....
(b) When no. 4 cyl. is on the exhaust stroke, no. 1 cyl. is on...............................

SIX-CYLINDER

The line diagram shows a six-cylinder in-line engine:

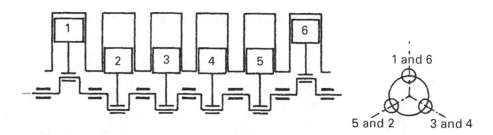

1 and 6

5 and 2   3 and 4

INVESTIGATION

Determination of firing orders on a four- and six-cylinder engine when the distributor is removed.

Remove the valve-rocker covers and chalk all the inlet valves. Turn the engine in the correct direction of rotation. The sequence in which the inlet valves open is the same as the engine firing order.

Vehicle checked.............................   Firing order   ....

Vehicle checked.............................   Firing order   ....

The cylinder build-up can be 2, 4, 6, 8 or 12 cylinders in two separate banks.

© ROVER

VEE ENGINES

© SUBARU

HORIZONTALLY OPPOSED ENGINES

Identify the number of cylinders on the engines below.

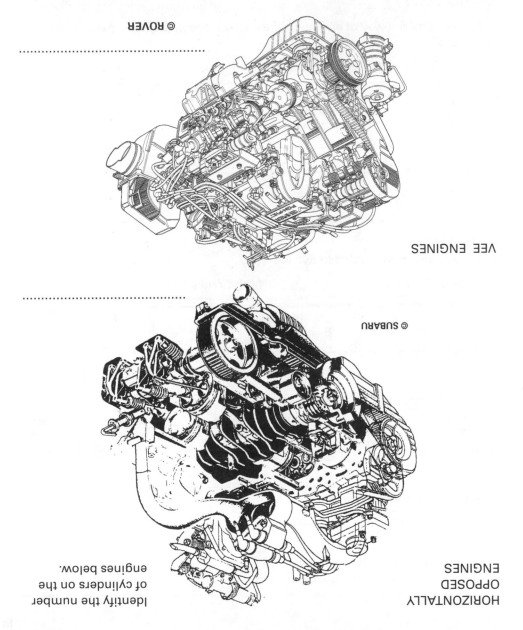

Define the meaning of firing order:

....

HORIZONTALLY OPPOSED

....

VEE

....

State the relative advantage of each cylinder arrangement:

....

IN-LINE

....

On each engine below, number the cylinders as quoted by engine manufacturers, and state their firing orders

	Vee 8	Vee 6	Vee 4	Horizontally opposed
Make				
Firing order				

COMBUSTION PROCESSES

Describe how combustion is made to occur in the combustion chamber of both the petrol and compression-ignition engine.

COMBUSTION OF A PETROL/AIR MIXTURE

....
....
....
....
....
....
....
....
....
....

COMBUSTION IN A DIESEL (COMPRESSION-IGNITION) ENGINE

....
....
....
....
....
....
....
....

CYLINDER HEAD – COMBUSTION CHAMBER SHAPES (SPARK-IGNITION)

Describe the three main types of combustion chambers shown, making reference to the position of the valves and spark plug.

SPARK-IGNITION DESIGN – PETROL   wedge

....
....
....
....
....
....

hemispherical

....
....
....
....
....
....

bowl in piston

....
....
....
....
....
....

COMBUSTION SEQUENCES (SPARK-IGNITION)

Normal combustion

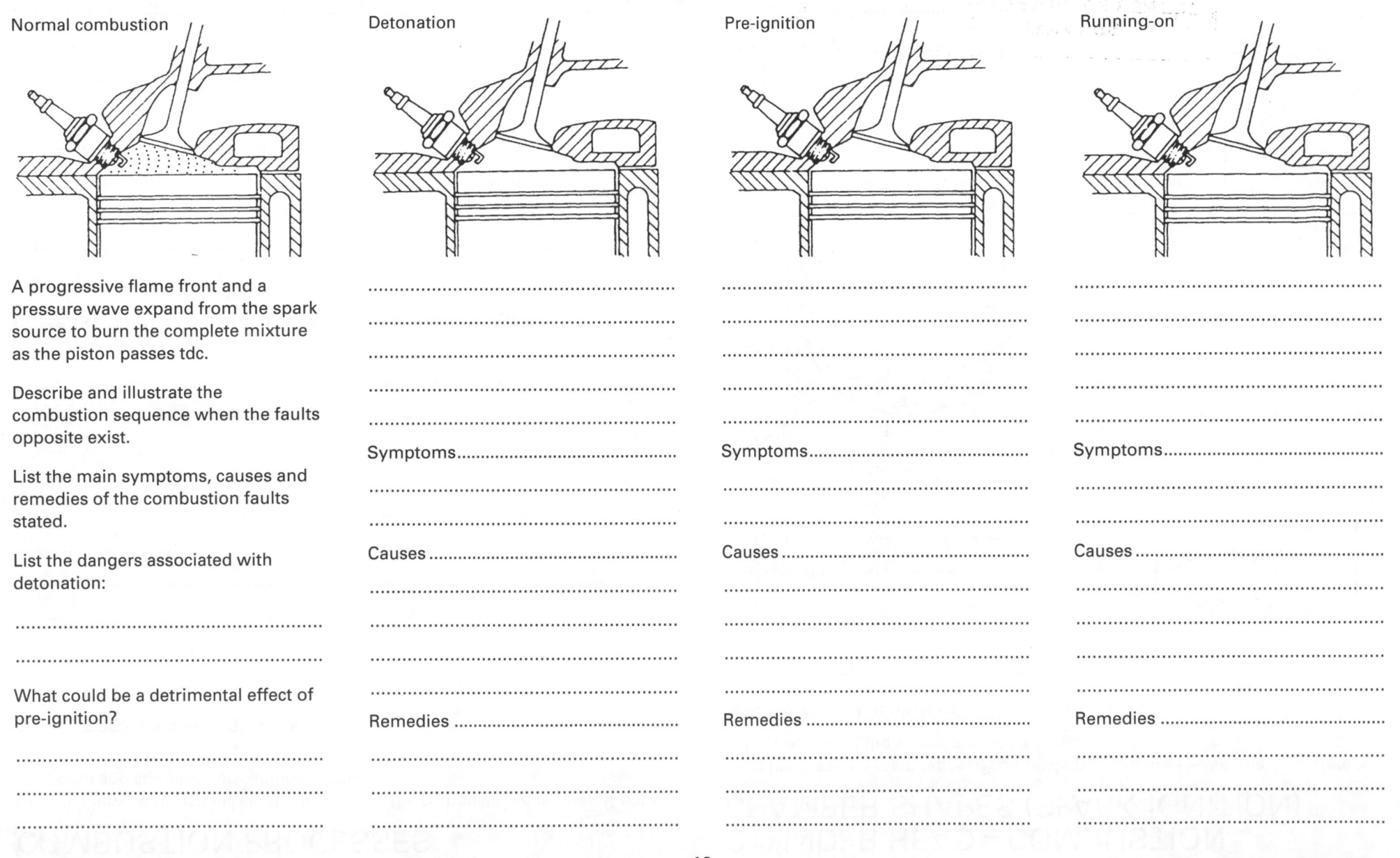

Detonation

Pre-ignition

Running-on

A progressive flame front and a pressure wave expand from the spark source to burn the complete mixture as the piston passes tdc.

Describe and illustrate the combustion sequence when the faults opposite exist.

List the main symptoms, causes and remedies of the combustion faults stated.

List the dangers associated with detonation:

....

....

What could be a detrimental effect of pre-ignition?

....

....

....

Detonation

....
....
....
....
....

Symptoms...
....
....

Causes ....
....
....
....

Remedies ....
....
....

Pre-ignition

....
....
....
....
....

Symptoms...
....
....

Causes ....
....
....
....

Remedies ....
....
....

Running-on

....
....
....
....
....

Symptoms...
....
....

Causes ....
....
....
....

Remedies ....
....
....

COMBUSTION CHAMBERS USED IN DIESEL (CI) ENGINES

DIRECT INJECTION

In the first diagram draw a sectioned view of the type of piston used in a direct injection engine.

Show on both diagrams the direction of air and fuel swirl as the piston is nearing top dead centre. Name the main parts.

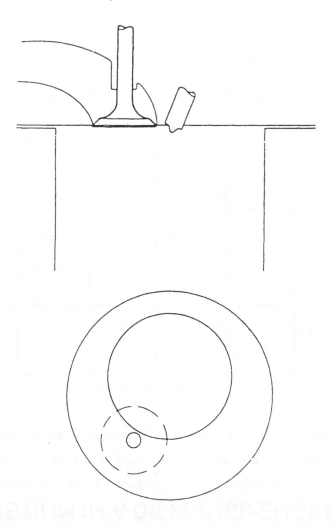

Describe the important features of the basic layout shown and explain how turbulence and swirl are induced and controlled:

....

....

....

....

....

INDIRECT INJECTION

The type of indirect injection shown is known as the Ricardo design.

PISTON CROWN

Name the main parts.

Explain how turbulence is created in the pre-combustion chamber shown:

....

....

....

....

COMBUSTION IN A DIESEL (CI) ENGINE

The combustion process can be described in three distinct stages or phases.

Complete the graph to show the approximate pressure rise in a CI engine.

Indicate the three phases of combustion and explain what is happening at the various stages.

Pressure crank angle diagram

Pressure

50°   25°   0°   25°   50°

btdc   tdc   atdc
(before top   (top dead   (after top
dead centre)   centre)   dead centre)

....

....

....

....

....

....

The first phase of combustion is known as the 'delay period'. What is meant by this term?

....

....

....

What affects the length of the delay period?

....

....

....

What is the effect on 'knock' when the delay period is varied?

....

....

....

The amount of fuel supplied and the serviceability of the injector play an important part on the performance output of an engine.

State how the engine performance would be affected by:

(a) excessive fuel

....

....

....

(b) poor atomisation

....

....

....

VALVE TIMING

The inlet valve as well as being open on the induction stroke is also partly open on two other strokes.

The inlet valve opens on the...

and closes on the..

The exhaust valve opens on the ....

and closes on the..

The reasons for these early openings and late closings are................................

....

....

What is meant by the following and when do they occur?

Valve overlap...

....

....

Valve lead ....

....

....

Valve lag ....

....

....

Indicate on the diagrams examples of where lead, lag and overlap occur.

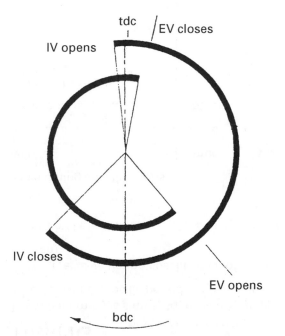

The valve-open period can be represented on a valve timing diagram. This shows the number of crankshaft degrees during which the valves are open.

The diagram left shows a valve timing diagram and indicates the open and close points.

Measure and state the angles before and after tdc.

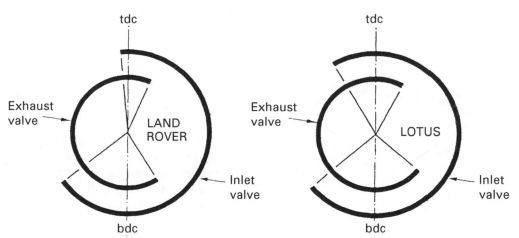

DIESEL (CI) FOUR-STROKE CYCLE – VALVE TIMING

The valve timing is similar to the spark-ignition engine timing, but since the fuel is injected over a specific period it too should be shown.

Obtain the valve timing specification of a CI engine and draw the valve timing diagram.

Vehicle make .... Model...

Engine type .... Capacity ....

Valve timing specifications

IVO .... btdc   EVO .... bbdc

IVC .... abdc   EVC .... atdc

Spill point.................................btdc

Total inlet valve   =
open period   ....

Total exhaust valve   =
open period   ....

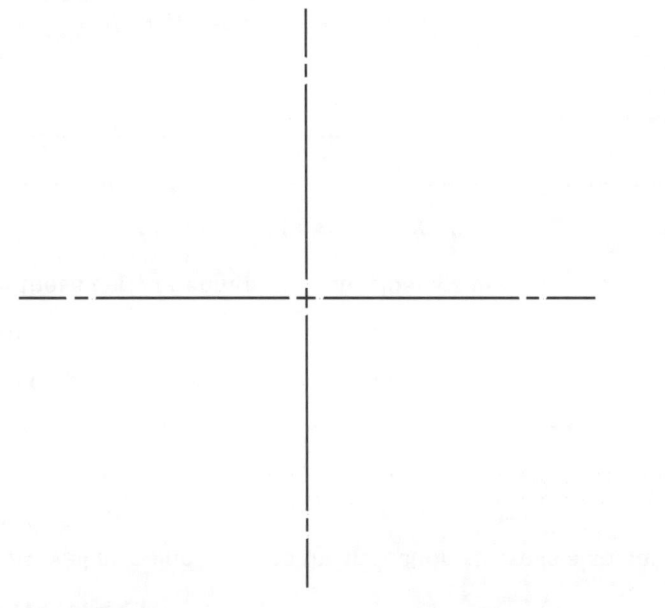

VALVE-PORT TIMING DIAGRAMS TWO-STROKE SPARK-IGNITION

All the operations occur in one crankshaft revolution.

Engines using ports only and employing the crankcase as an induction pressure chamber always have equal port openings either side of tdc or bdc.

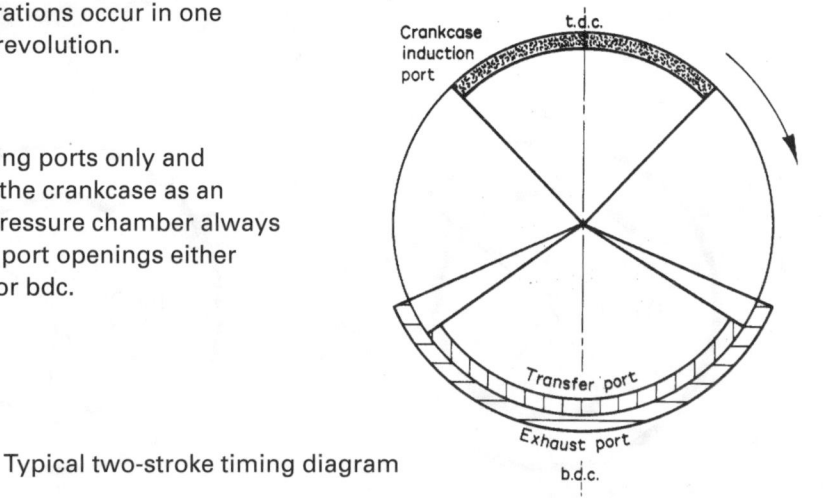

Typical two-stroke timing diagram

TWO-STROKE COMPRESSION-IGNITION

These have a similar diagram but without the transfer port at the top.

Obtain the valve timing specifications for a two-stroke CI engine using INLET PORTS AND EXHAUST VALVES and complete the diagram and table below.

Vehicle make ....   Model ....

Engine make ....

Capacity ....

IPO ....

IPC ....

EVO ....

EVC ....

Spill point ....

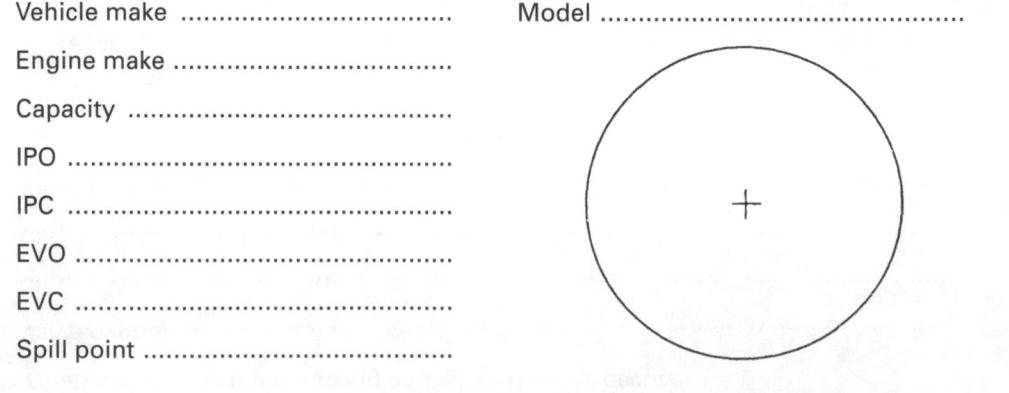

Chapter 2

Engine Lubrication

Engine lubrication systems	18	Oil pressure warning light switch	25
Purpose of a lubricant	19	Oil filters	26
Properties of oil	19	Oil cooler, pipes and hoses	27
Viscosity	20	Oil level indicator	27
Viscosity index	20	Crankcase emission control	27
Engine oil SAE viscosity classification	20	Crankcase ventilation valve	28
Multigrade oils	21	Cylinder-head gaskets	29
Wet sump system	22	Types of oil-sealing arrangements	29
Dry sump system	22	Engine lubrication system – protection during use and repair	30
Oil pumps	23	Maintenance	31
Oil pressure relief valves	24	Equipment necessary for repair	31
Oil pressure gauges	25	Diagnostics	31

ENGINE LUBRICATION SYSTEMS

In all modern four-stroke engine lubrication systems the principal bearing surfaces, with the notable exception of the pistons, are supplied with oil under pressure by a positively driven high-pressure pump; other parts are lubricated by splash or mist.

Name the parts which make up the engine lubrication system.

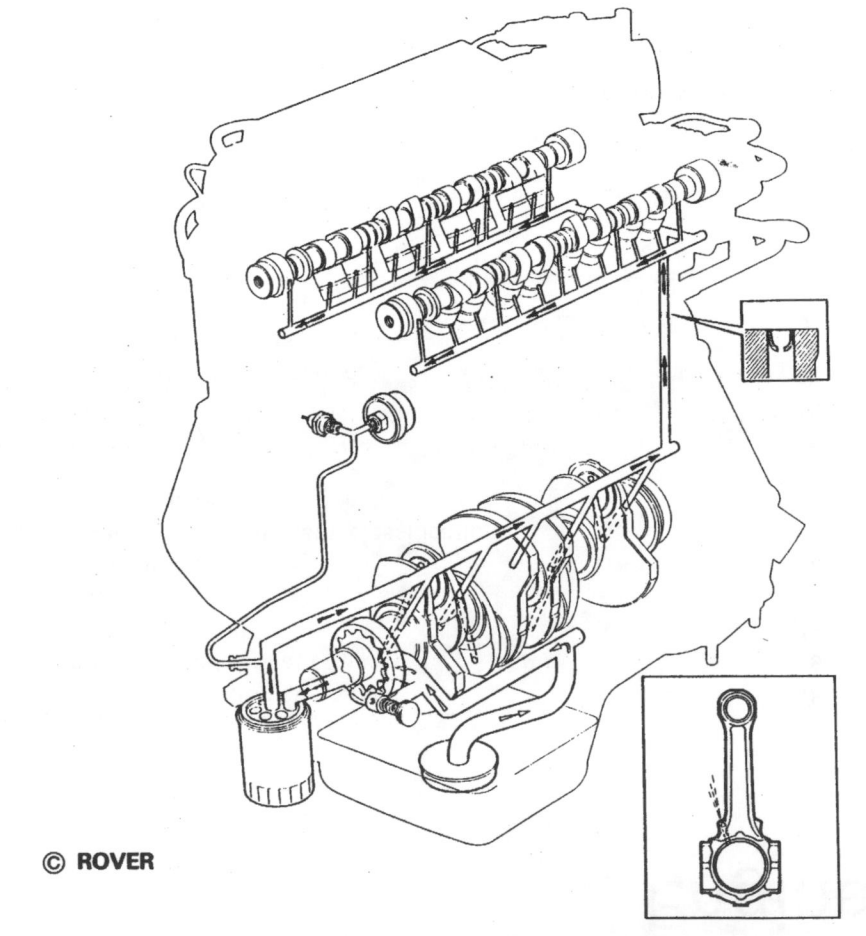

© **ROVER**

The lubrication system shown is known as

....

With the aid of the diagram describe how the oil is distributed by pressure, splash and mist.

....

....

....

....

....

....

....

....

....

....

....

....

How may pistons be better cooled in high performance engines?

....

....

....

Lubrication systems may be classified into three systems. These are:

1. ....   2. ....   3. ....

The first two types are covered in the following pages, the third type is only used in small two-stroke petrol engines.

What is meant by 'total loss'?

....

....

....

....

PURPOSE OF A LUBRICANT

All moving components usually have some form of lubricant placed between the surfaces.

The purpose of a lubricant is to: ....
....
....

The types of lubrication may be classed as:

1. ....   2. ....

Lubrication should be based on maintaining a fluid film whenever possible and it is the behaviour of an oil under extreme load and temperature conditions which determines the usefulness of an oil.

The two drawings below each show a much-magnified representation of part of a bearing face. Add a similar sketch above each one, in such a way as to show what is meant by 'full-fluid film' and 'boundary' lubrication.

Full-fluid film lubrication   Boundary lubrication

When only boundary lubrication is present sometimes the layer of oil is only one molecule thick and could easily break down.

What would occur if the film of lubricant broke down, because of the high temperatures and pressures that occur in the engine?

....

PROPERTIES OF OIL

Oil possesses two main properties: body and flow. Briefly explain their importance.

Body is the ability of an oil to maintain an oil film between two surfaces.

....
....
....
....
....
....
....

Flow concerns the property of an oil to spread easily over surfaces and to flow through pipelines and oilways.

....
....

How does the change in temperature affect these two properties?

....
....

Modern oils use many additives to improve the above and other properties. List the requirements that a modern engine oil must meet.

....
....
....
....
....
....
....

VISCOSITY

The viscosity of an oil is a measure of its thickness (or body), or (more correctly) of its ....

This property varies with temperature and for a specific oil:

When the temperature is low the oil will be ....

When the temperature is high the oil will be ....

An oil said to be 'thin' is more properly described as a ....

.... and an oil described as 'thick' should be called a

....

The viscosity of an oil is measured using a viscometer. This measures

....

....

VISCOSITY INDEX

It is important that oil viscosity remains as stable as possible during changes in temperature. Define viscosity index.

....

....

....

....

....

....

....

....

ENGINE OIL SAE VISCOSITY CLASSIFICATION

The viscosity of an oil is expressed as a number prefixed by the letters SAE, for example SAE 30.

SAE stands for the .... which is the American organisation that devised these viscosity standards.

State TWO typical engine oil viscosity numbers:

1. .... 2. ....

Which oil has the lower viscosity? ....

Modern oils often have two viscosity numbers, and are called

....

The reason for two numbers is that two measuring standards are used, one at engine working temperature and the other at a very low temperature.

The working viscosity is calculated with the oil at 99°C (210°F) and the grade of oil is expressed as:

SAE 20, 30, 40, 50 etc.
The most viscous of the above is...

The second viscosity range is calculated at − 18°C (0°F).

These are very low viscosity oils and have a suffix W (Winter) to indicate the measuring standard, for example:

SAE 5W, 10W, 20W.

The lowest viscosity there is...

A modern multigrade oil is an oil whose viscosity meets the flow standards measured at both temperatures. This has many advantages:

....

....

....

....

Note: the SAE number signifies only the viscosity of the oil at the specified temperature. It in no way indicates the quality of the oil.

MULTIGRADE OILS

By improving the viscosity index of an oil (using suitable additives) it is possible for the oil to fall into two viscosity ranges when tested.

State TWO typical multigrade viscosities: ....

Why, if possible, should a 10W30 oil be used in preference to a 20W50 oil?

....

....

....

LUBRICATING OIL (TYPE AND GRADE)

When selecting oil for an engine, it is important that as well as being of the correct grade it must be the correct type to make it suitable for a particular application. What is the difference between grade and type?

....

....

....

....

....

How does a diesel engine oil differ from a petrol engine oil?

....

....

....

....

What type of oil does a turbocharged engine require?

....

....

....

....

SYNTHETIC OIL

A conventional multigrade oil will lose its hot grade viscosity and therefore its lubrication protection as the engine mileage increases. A 20W50 grade oil may reduce to a 20W30 grade oil after 6000 miles' engine use. This is because the long-chain viscosity improver additive, which maintains the oil's thickness when hot, gets chopped up by the shearing action of the moving parts.

A synthetic oil is created from crude oil by being refined in a more complex manner. The crude is put through a cracking process which produces oil having smaller molecules. This makes the oil more shear stable at high temperatures and so the normal VI improver is not required. A grade example would be 5W50 VI 185.

What are the advantages of using synthetic oil?

1. ....

2. ....

3. ....

OIL CONTAMINATION

Earlier in this chapter it was mentioned that the engine oil is subjected to extremely high pressures and temperatures. This affects both its viscosity and oiliness. The engine oil also has to contend with other problems, the main one being contamination.

List below the principal contaminants and their source or causes.

Contaminant	Source or cause
....	....
....	....
....	....
....	....
....	....
....	....

Why is it necessary to change engine oil at regular intervals whereas other components are filled for life?

....

....

WET SUMP SYSTEM

The thick black lines indicate the oil passageways and bearing surfaces. Add names to the main components of the system.

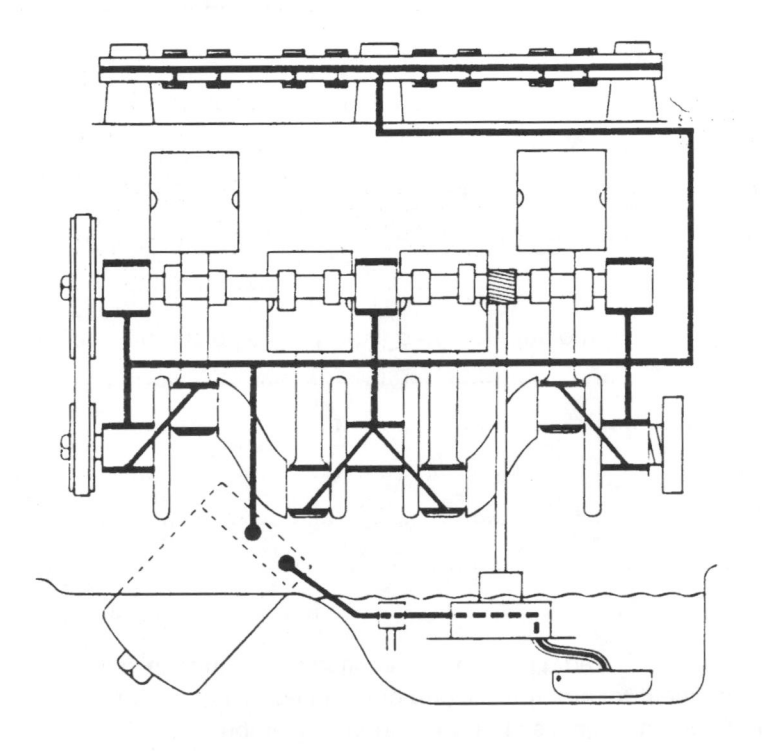

Filtration Systems

There are two types of filtration systems. These are:

.... and ....

The system shown above is a .... type.

Describe what is meant by a full-flow system.

....

....

....

DRY SUMP SYSTEM

Using a coloured crayon show the oil passageways and bearing surfaces. Indicate direction of oil flow.

Describe below the system's basic operation.

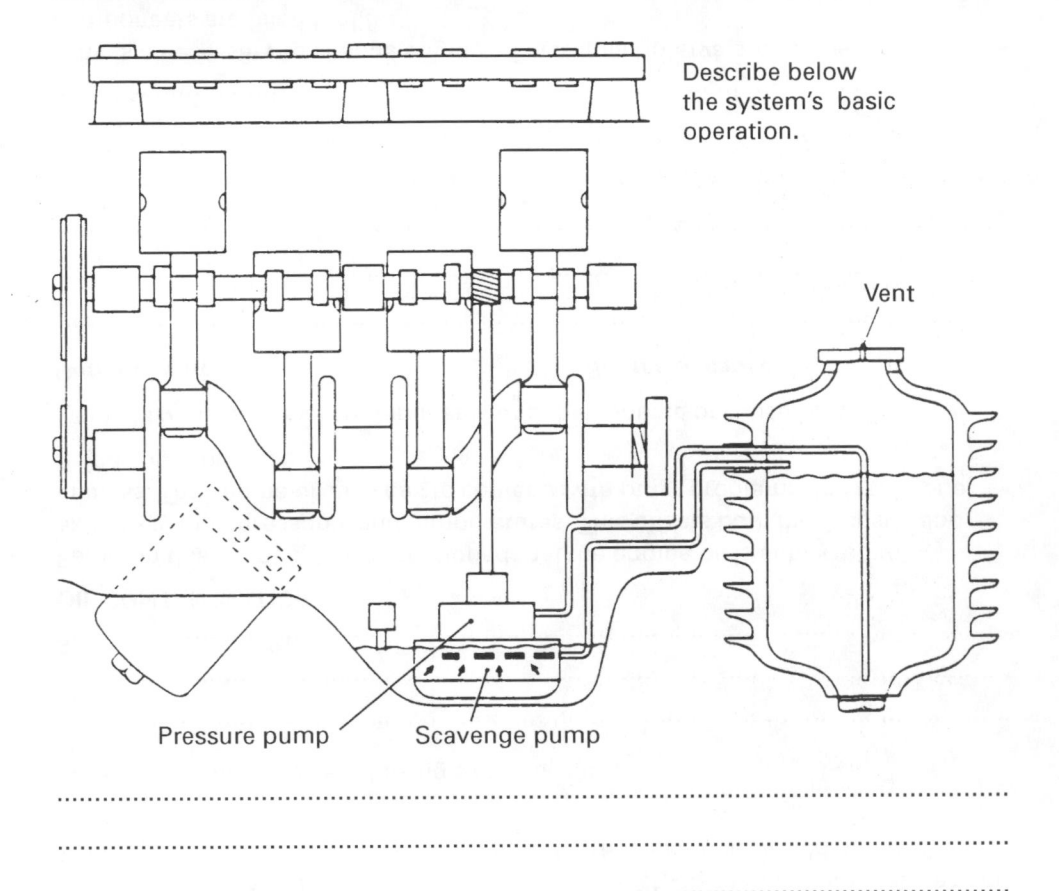

Vent

Pressure pump   Scavenge pump

....

....

....

....

What are the advantages of the dry sump system over the wet sump system?

....

....

22

OIL PUMPS

Examine various oil pumps similar to the ones shown below. On the sectional drawings show the direction of oil flow and pump rotation. Briefly describe the action of each pump and name the parts indicated.

Eccentric rotor   **Gear type**   **Eccentric vane**

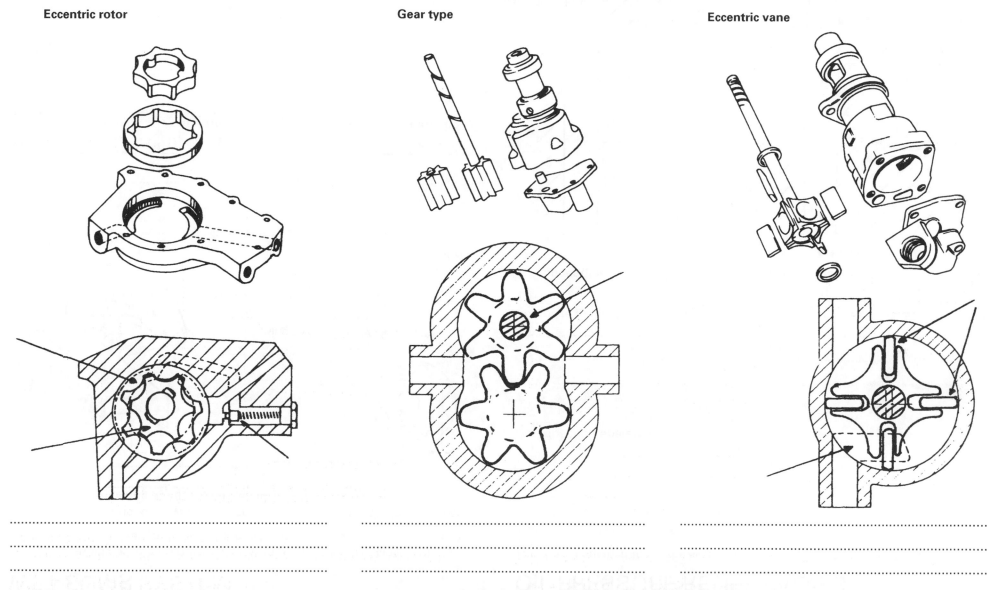

OIL PUMPS

In order to supply oil at high pressure the pump must maintain close working tolerances. This is even more critical when the pump is not submerged in the oil.

Examine an eccentric rotor-type pump and measure the working clearances as shown to determine the pump's serviceability.

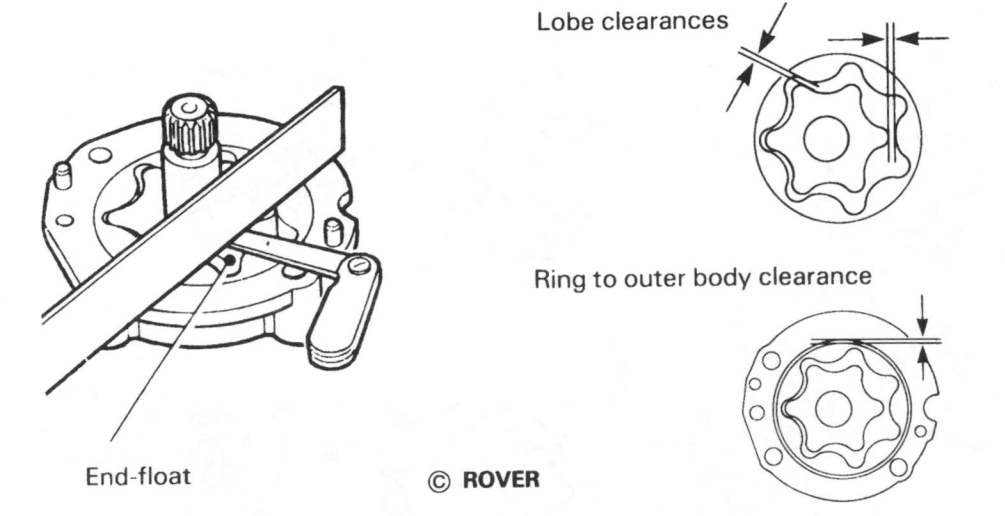

Lobe clearances

Ring to outer body clearance

End-float

© ROVER

Vehicle make .... Engine capacity ....

Clearance (mm)	End-float		Rotor lobe	Ring to outer body
	outer ring	rotor		
Manufacturer's specification				
Measurement taken				

COMMENTS ....
....

OIL PRESSURE RELIEF VALVES

These are usually fitted in the system between the oil pump and the oil filter.

Why is it necessary to fit a relief valve?

....

....

....

Describe their operation:

....

....

....

....

Complete the first sectioned sketch to show a plunger type oil pressure relief valve.

1. Plunger type

2. Ball type

To filter

From pump

From pump   Oil return to sump

OIL PRESSURE GAUGES

The gauge below is of the type designed to give a direct oil pressure reading.

This gauge is known as a Bourdon Tube type. Name the main parts.

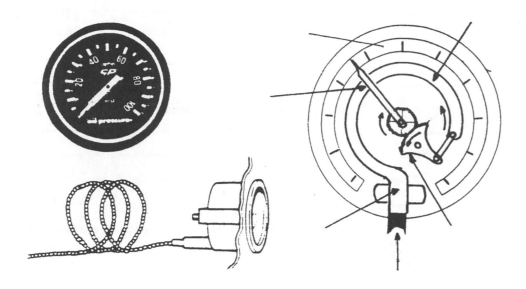

OPERATION

....

....

An alternative type of gauge would operate electrically in a manner similar to a fuel gauge.

Quote two vehicles that fit a pressure gauge as standard and state typical pressures.

Make	Model	Oil Pressure

OIL PRESSURE WARNING LIGHT SWITCH

This is the most common oil pressure indicator arrangement used on mass-produced vehicles.

The sectioned view shows the internal components of an oil pressure switch. Name the main parts and explain how it operates.

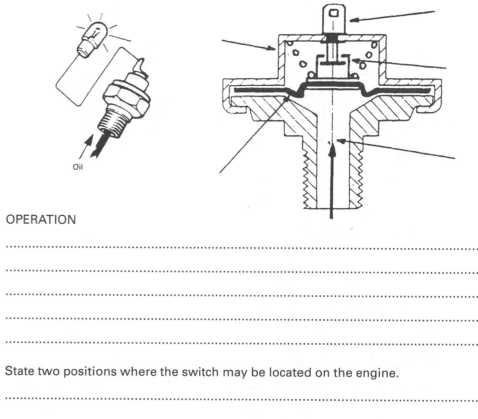

Oil

OPERATION

....

....

....

....

State two positions where the switch may be located on the engine.

....

....

Investigation

Connect pressure gauge to engine and note pressure when light goes out. Compare with pressure stamped on switch.

Switch operating pressure	Manufacturer's specification

OIL FILTERS

Most engines are fitted with two filters. The first (primary) filter is fitted around the pump intake in the sump and is of a coarse wire mesh type. The second (secondary) filter is of the replaceable element type.

State the purpose of each filter:

Primary filter...

....

Secondary filter ....

....

What is the purpose of the full-flow filter by-pass valve?

....

....

....

Describe the basic construction of the filter element:

....

....

....

Which way does the oil flow through the element?

....

Describe how the filter should be removed and refitted:

....

....

....

....

The diagram below shows a sectioned view of a full-flow oil-filter element and the by-pass valve. Name the main parts and show the direction of oil flow.

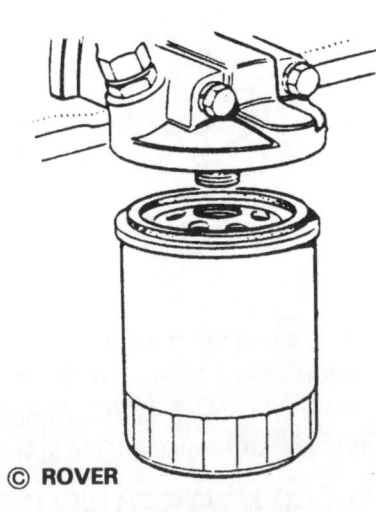

© ROVER

26

OIL COOLER, PIPES AND HOSES

Many high-performance engines, because of very high operating temperatures, require extra oil cooling. This they achieve by use of a small external radiator similar to the one shown below.

Describe how the oil flow is controlled through the radiator:

....
....
....
....
....
....
....

From engine

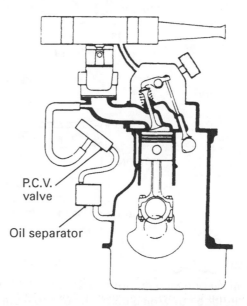

Describe the type of oil pressure hoses used:

....
....

OIL LEVEL INDICATOR

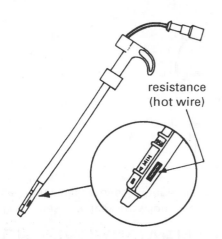

resistance (hot wire)

The most popular oil level indicator is a 'hot wire' dip stick as shown below.

The dip stick is a hollow plastic moulding. A resistance wire is placed inside at a position between the oil level marks.

Describe how this is able to test the oil level.

....
....
....
....
....

CRANKCASE EMISSION CONTROL

On modern cars it is usual to employ some form of crankcase ventilation which does not emit unburnt oil fumes to the atmosphere. What is the reason for this practice?

....

Give two reasons why a flow of air through the crankcase is necessary:

....
....
....
....
....
....
....

The diagram shows a positive crankcase ventilation (PCV) system of the fully closed type. Add arrows to indicate ventilation flow.

P.C.V. valve

Oil separator

27

CRANKCASE VENTILATION VALVE

Two types of valves are widely used. They are both controlled by manifold vacuum which moves either a diaphragm or a plunger (as shown below). Label the sketch.

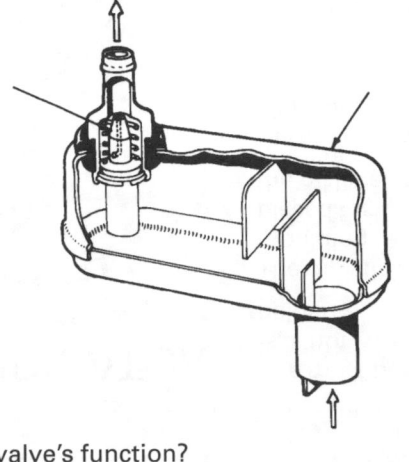

What is the ventilation valve's function?

....

....

....

....

What disadvantages may the valve cause?

....

....

....

....

How should these valves be serviced? State a typical mileage period.

....

....

Explain, with the aid of the diagram below, the operation of the A–C crankcase ventilation valve, giving reasons for the flow and no-flow positions.

1. Engine off or backfire

© A-C DELCO

Plunger in closed position

No manifold vacuum

No flow

2. Idling or low speed

Plunger in seated position

High manifold vacuum

Minimum flow

3. High speed

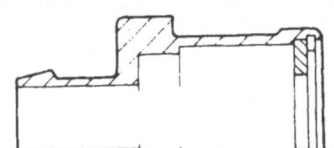

Show the position of the plunger, and the flow of gas, when the engine is rotating at high speeds.

....

....

....

....

....

....

....

....

....

....

....

....

....

....

....

CYLINDER-HEAD GASKETS

Identify the hole uses:

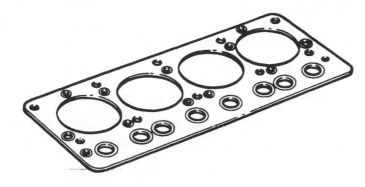

State the purpose of a cylinder-head gasket:

....

....

....

....

Give some of the main symptoms and causes of cylinder-head gasket failure:

Symptoms

....

....

....

....

Causes

....

....

....

....

TYPES OF OIL-SEALING ARRANGEMENTS

Identify the types of engine oil-sealing arrangements shown and label the parts indicated:

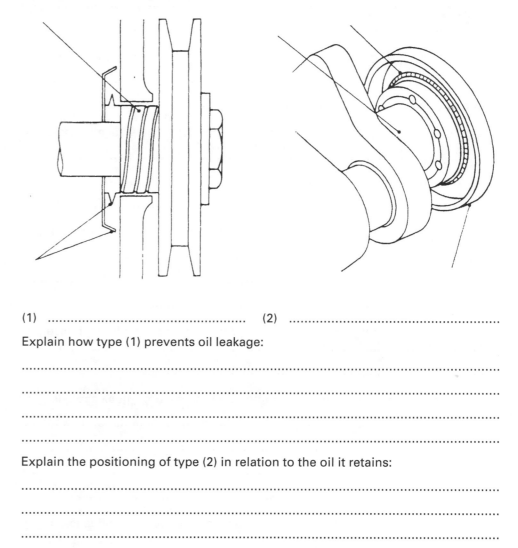

(1) .... (2) ....

Explain how type (1) prevents oil leakage:

....

....

....

....

Explain the positioning of type (2) in relation to the oil it retains:

....

....

....

....

ENGINE LUBRICATION SYSTEM – PROTECTION DURING USE AND REPAIR

Describe how the system should be protected from the following hazards:

1. Ingress of dirt ....
....
....
....

2. Crankcase dilution ....
....
....

3. Crankcase contamination ....
....
....
....
....

4. Overheating ....
....
....

5. Excess pressure ....
....
....

How should the following be maintained/checked to protect the lubrication system and its components?

1. Oil filter gauze in sump ....
....

2. Correct engine operating procedures ....
....
....
....

3. Crankcase ventilation ....
....
....

4. Sealing ....
....
....

5. Use of oil ....
....
....

6. Pressure relief valve ....
....
....

7. Oil coolers ....
....

8. Filters ....
....

MAINTENANCE

List five main precautions to be observed when carrying out routine maintenance:

....

....

....

....

....

....

EQUIPMENT NECESSARY FOR REPAIR

State six of the main tools required to carry out work on engine lubrication systems:

....

....

....

....

....

DIAGNOSTICS: ENGINE LUBRICATION – SYMPTOMS, FAULTS AND CAUSES

State a likely cause for each symptom/system fault listed below. Each cause will suggest any corrective action required.

SYMPTOM	PROBABLE FAULT	LIKELY CAUSE
The oil warning light comes on when engine is warm.	Low oil pressure	....
Oil seems to be leaking from the engine seals and particularly from the crankshaft.	High oil pressure	....
Blue smoke is coming from the engine's exhaust after starting when cold.	High oil consumption	....
There are oil drops on the floor every time the vehicle is parked.	Oil leakage	....
The oil is very black when checked.	Oil contamination	....
The oil level on the dip stick is higher than normal.	Overfilling Water in oil	....

Chapter 3

Cooling and Heater Systems

Cooling systems	33	Water heaters	43
Air-cooling system	34	Hoses	44
Water-cooled pressurised system	35	Shutter controls	44
Alternative liquid-cooling system layouts	36	Effects of freezing	45
Thermostat	37	Antifreeze	45
Pressure cap	38	Effects of corrosion – acids and alkalis	46
Radiator and heater matrix	39	Cooling system protection during use	46
Fans	41	Diagnostics	47
Water pumps	42	Routine maintenance	48
Sealed cooling systems	42	Heat loss calculations	49

COOLING SYSTEMS

The principle of the heat engine, which derives its power from the combustion of fuel and air and the resultant expansion of gases, necessitates the use of some type of cooling system.

During combustion the temperature in the cylinder can momentarily be as high as 1800°C. Even when the gases expand and the temperature falls, it may still be higher than the melting point of aluminium.

What problems may occur if this heat is not dissipated at the correct rate?

....

....

....

....

The two main types of cooling systems are AIR cooled and LIQUID cooled. List TWO subdivisions of each type:

AIR COOLING

1. ....

2. ....

LIQUID COOLING

1. ....

2. ....

State FOUR functional requirements of a cooling system:

1. ....

2. ....

3. ....

4. ....

....

Heat Transmission

Heat is one type of energy. A piston engine relies on heat for its principle of operation. Heat is transmitted in one (or more) of three ways.

....

....

....

....

....

....

....

....

The sketch below shows part of an air-cooled engine. Add arrows showing the direction the heat path takes from the piston crown to the outside of the cylinder.

Heat flow, through the metal is by ....

AIR-COOLING SYSTEM

With a few notable exceptions, this system is not very popular for multi-cylinder engines as difficulties are encountered trying to cool equally all the cylinders and maintain a constant temperature.

How is the air flow normally controlled?

....

....

....

....

....

....

Name the main parts indicated on the air cooled FIAT engine shown below:

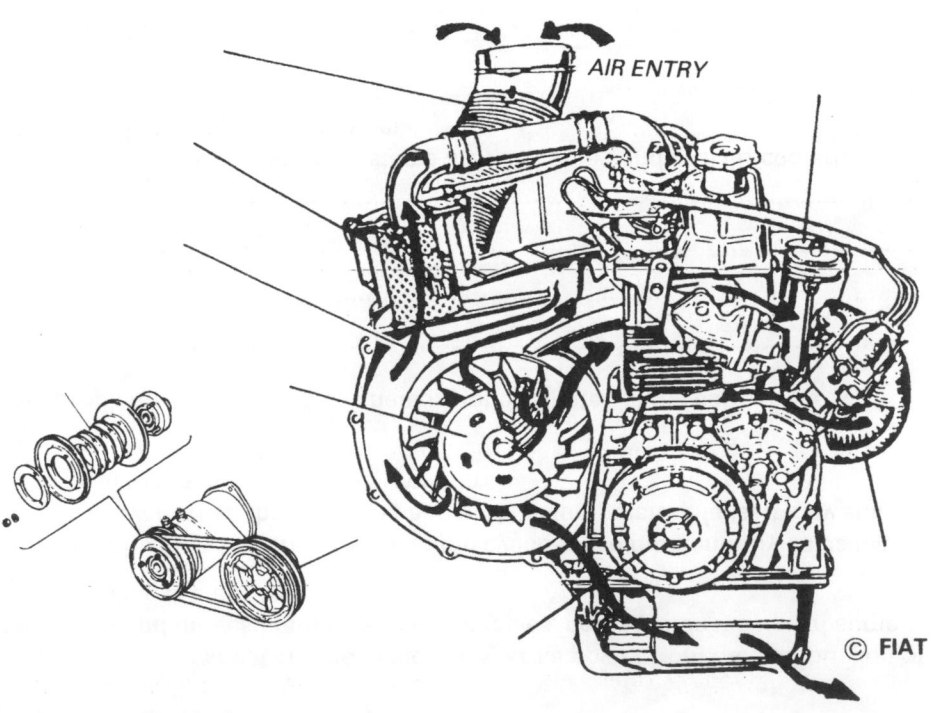

AIR ENTRY

© FIAT

Indicate the air flow through the horizontally opposed engine shown below, and name important parts:

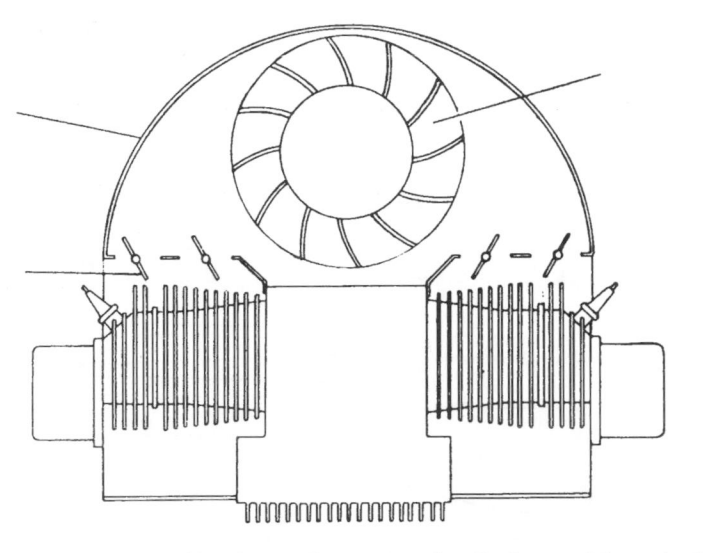

What are the advantages of horizontally opposed cylinders, with regard to air cooling?

....

....

Why is it particularly important on an air-cooled engine that fan-belt tension is correct?

....

....

....

....

The air fan is commonly driven from the rear of the generator and fan-belt tensioning is adjusted by means of a split pulley.

A   B

Complete the sketches to show how the belt is tensioned.

AIR-COOLING SYSTEM

What feature controls the rate of air flow over the engine cylinders?

....

....

....

....

Describe the interior car heater system layout and the method by which the hot air is produced when the engine is air cooled:

....

....

....

....

....

....

Advantages of Air and Liquid Cooling Systems

State the relative advantages of liquid cooling compared with air cooling:

....

....

....

State the relative advantages of air cooling compared with liquid cooling:

....

....

....

....

WATER-COOLED PRESSURISED SYSTEM

Water is a better cooling medium than air; it has a high specific heat and is able to transfer heat more efficiently.

Unfortunately for water-cooling systems, the engine gives its best thermal efficiency when the cooling water is close to 100°C (that is, the normal boiling point). To overcome this problem water-cooling systems are normally pressurised.

A cooling system is pressurised to ....the .... temperature of the coolant.

Why is such a design feature considered necessary?

....

....

What advantages are gained by pressurising the system?

....

....

....

....

....

....

....

What limits the pressure that can be imposed on a water-cooling system?

....

For approximately every 5 kPa of pressure rise the water boiling temperature increases by 1°C.

What would the approximate boiling temperature of water in a cooling system be, using:

(i) a 4 lbf/in^2 (30 kPa) pressure cap? ....

(ii) a 15 lbf/in^2 (105 kPa) pressure cap? ....

ALTERNATIVE LIQUID-COOLING SYSTEM LAYOUTS

The liquid-cooling system basically consists of water jackets surrounding the cylinders with provision for the heated water to pass into a radiator and cooled water from the radiator to flow back into the cylinder block.

Using arrows indicate the flow of water through the systems shown and name the main parts or features.

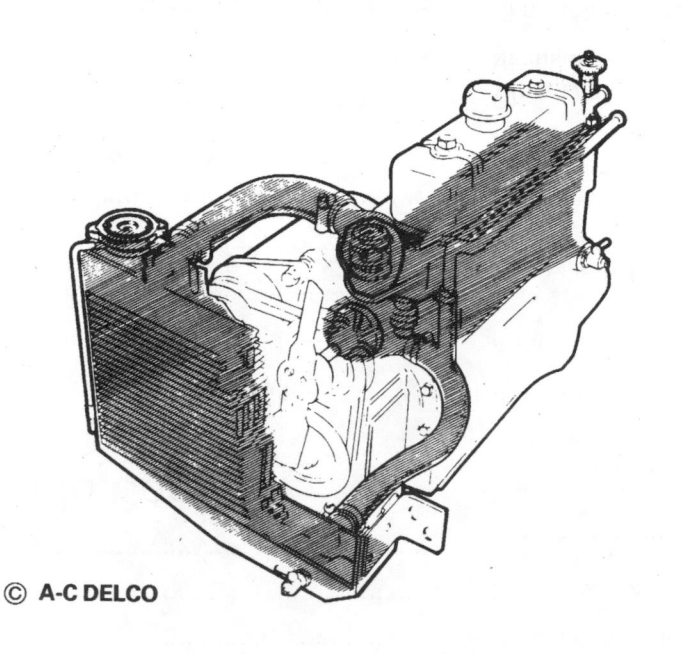

© A-C DELCO

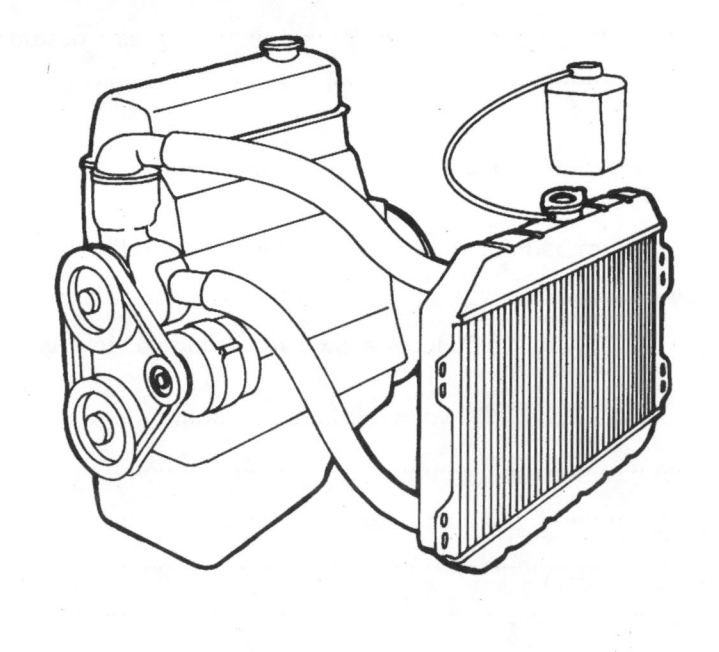

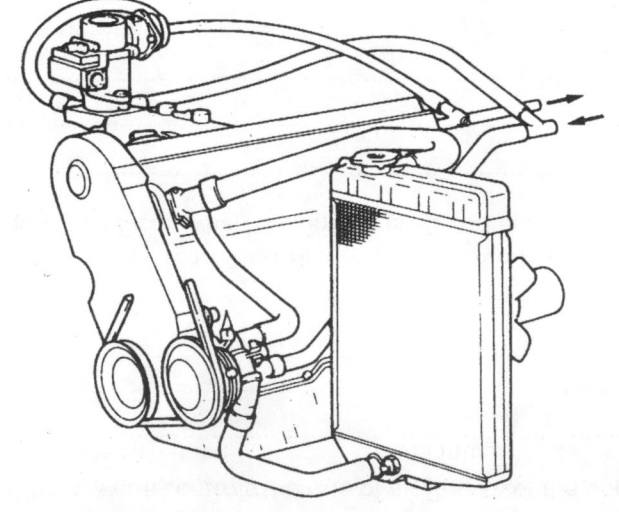

© VOLKSWAGEN

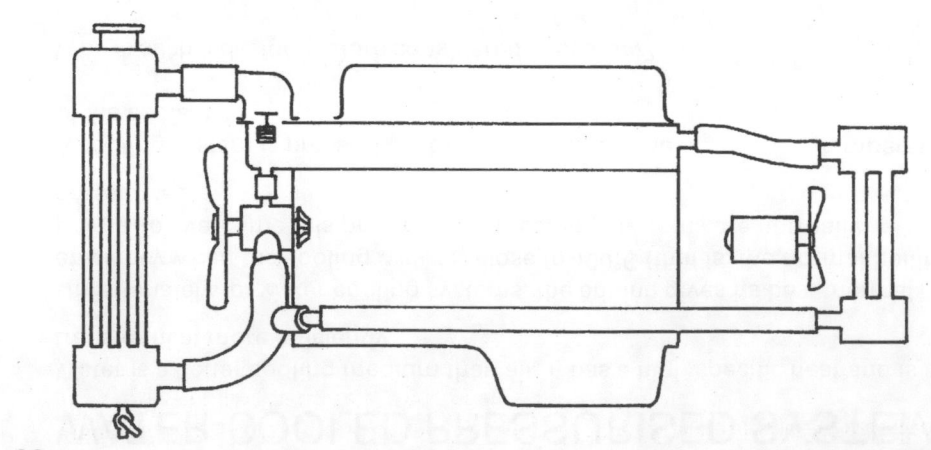

THERMOSTAT

This is a temperature-sensitive valve that controls water flow to the radiator. State the two main reasons why it is fitted:

1. ....

....

2. ....

....

Wax Element Type

This type employs a special wax contained in a strong steel cylinder into which passes the thrust pin. This is surrounded by a rubber sleeve which also seals the upper end of the cylinder.

The wax element type thermostat is shown fully closed on the left and fully open on the right. Show the water flow direction on the right-hand sketch and name the main parts.

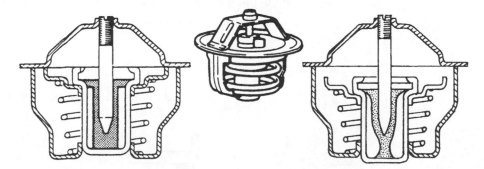

Describe its basic operation ....

....

....

....

State the purpose of the small hole and pin in the valve disc:

....

....

All substances exist in one of three states: solid, gas or liquid. Most can change from one to the other. State a common example.

....

....

What change of state causes the wax type thermostat to operate?

....

Thermostat Operation Test

Check the serviceability of a thermostat when it has been removed from the vehicle.

Sketch below the equipment used:

Thermostat make.. Type ....

Specified opening temperature ....

Actual temperatures: opening .... closing ....

Visual defects (if any) ....

....

Serviceability ....

PRESSURE CAP

This consists of a spring-loaded valve which resists the pressure of the expanding coolant, air and steam, in the header tank unit.

Name the essential parts of this typical radiator cap.

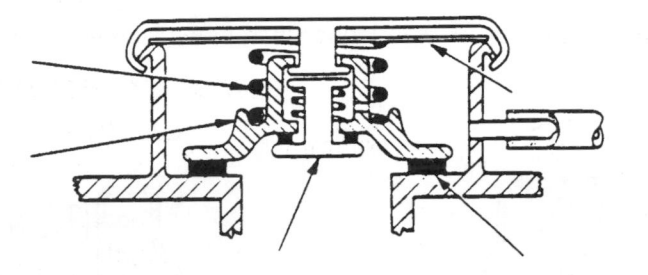

How is the pressure controlled?

....

....

....

Explain the purpose of the small valve fitted in the centre of the pressure cap:

....

....

Pressure Cap Testing

Connect cap to adapter and tester. Pressurise to check when valve opens. The obtained pressure should be:

....

....

....

ADAPTER

CAP

State what basic checks are being made in the sketches opposite:

1. ....

....

....

....

....

....

....

....

2. ....

....

....

....

....

....

....

Examine various caps and state at what pressure they should operate.

Use a pressure tester to check their condition.

Make of vehicle	Stated pressure of cap tested	Actual release pressure	State if cap maintained pressure for over 10 seconds

38

RADIATOR AND HEATER MATRIX

The purpose of the radiator is to provide a large cooling area for the water and expose it to the air stream. A reservoir for the water is also included in the construction. This is known as the header tank and is often made from thin steel or brass sheet. The header tank is connected to the bottom tank by numerous brass or copper tubes surrounded by cooling 'fins', and this assembly is known as the matrix, block, stack or core.

The matrix on some modern radiators is made of aluminium, using plastic for the header and bottom tanks.

Name the main parts.

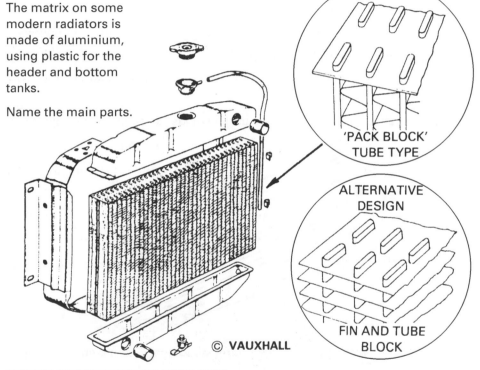

'PACK BLOCK'
TUBE TYPE

ALTERNATIVE
DESIGN

© VAUXHALL

FIN AND TUBE
BLOCK

MATRIX CONSTRUCTION-TUBE TYPE

This consists of thin, almost flat or oval copper or brass tubes arranged in rows. Why are oval tubes used?

....

....

....

Radiator Testing

PRESSURE TESTING

A pressure test is carried out to determine external or internal leaks.

Describe how to pressure test a system.

....

....

....

....

....

....

....

....

....

© CHRYSLER

Pressure testing the cooling system

FLOW TESTING

If a radiator is suspected of being partially blocked how should it be checked when:

1. It is on the vehicle?

....

....

....

2. It has been removed from the vehicle?

....

....

....

....

39

Cross-flow Radiators

In the conventional radiator the coolant flows ....

through the core from ....to ....

In cross-flow radiators the coolant flows ....

through the core from the .... of one side tank across to the

.... of the other side tank.

Indicate the direction of water flow through the radiator shown.

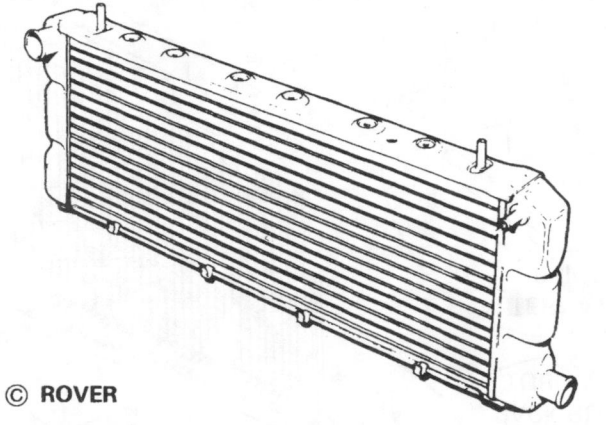

© **ROVER**

Why is it considered necessary to fit such a radiator in preference to the vertical flow type?

....

....

....

What are the cross-flow radiator's basic disadvantages?

....

....

....

....

With most cross-flow systems it is necessary to fit a remote header or expansion tank as shown below. Label the major parts on this drawing.

ROVER
V6
ENGINE

© **ROVER**

Why is the remote header tank necessary?

....

....

....

Air Bleeding

Describe a typical coolant filling procedure for a system that uses a remote header tank similar to the system shown above:

....

....

....

....

....

....

FANS

Most cooling systems (water or air) are fitted with fans. In the simplest arrangement the fan is permanently driven from the crankshaft via the fan belt.

The fan's function is: ....

....

Why is the type of fan described above rarely used on modern vehicles?

....

....

Viscous Coupling Fans

These units operate on the viscous shear principle.

There are TWO types of couplings:

1. TORQUE LIMITING. Here the thickness of the fluid determines the slipping effect and therefore the maximum speed at which the fan will rotate (3000 rev/min).

2. AIR TEMPERATURE SENSITIVE. In this type warm air acts on a bi-metal strip sensor; this sensor opens a valve which allows fluid into the clutch and so provides maximum drive.

What is indicated on the viscous coupling shown?

....

....

....

State the function of the fins on the viscous coupling:

....

....

....

Electrically Driven Fan

The construction of this type includes an electric motor complete with fan, a temperature-sensitive control unit and a warning light.

Name the parts indicated.

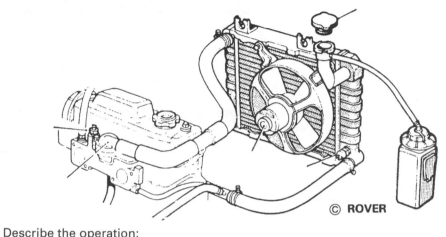

© **ROVER**

Describe the operation:

....

....

....

....

....

Describe how a thermostatically controlled cooling fan should be checked for correct operation:

....

....

....

....

....

....

WATER PUMPS

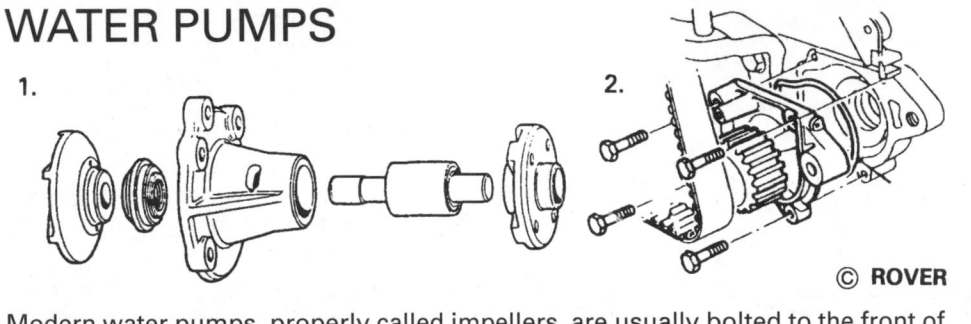

1.

2.

© **ROVER**

Modern water pumps, properly called impellers, are usually bolted to the front of the cylinder block and belt-driven from the crankshaft.

Examine various designs of water pumps and name the indicated parts shown on the sectioned view below:

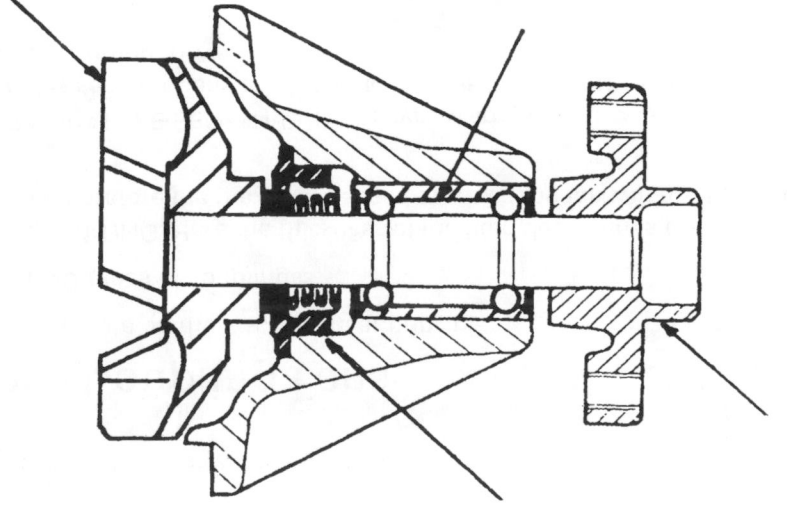

Describe the construction of the pump shown.

....

....

....

....

Explain the provision normally made to circulate the water expelled from the water pump when the thermostat is closed:

....

....

....

SEALED COOLING SYSTEMS

A disadvantage of the ordinary cooling system is that small losses of coolant occur through the radiator overflow pipe. If the level of water in the header tank is not frequently checked it is possible for the water level to fall sufficiently to prevent circulation. A method of overcoming this is simply to immerse the lower end of the overflow pipe in coolant contained in an expansion chamber.

A sealed system ensures that the cooling system is maintained completely full at all times. What design feature does this require?

....

....

Why is a sealed system considered essential on some heavy vehicles and PSVs?

....

....

List the advantages and therefore reasons for use of a sealed cooling system:

....

....

....

List the disadvantages of a sealed cooling system:

....

....

....

WATER HEATERS

To give adequate comfort to the occupants of a vehicle in varying weather conditions, all modern vehicles are equipped with some form of heating and ventilation system. The heat source, on vehicles fitted with water-cooled engines, is usually the hot water from the engine. Heater units which utilise the engine cooling water are normally situated on the bulkhead.

Describe the basic action of the heaters shown.

....

....

....

....

....

....

Drawing shows a fresh-air cab heater with its operational flaps in the off position.

Show the flaps positioned to give:

(i) Maximum heat to interior and screen

(ii) Warm air to screen only

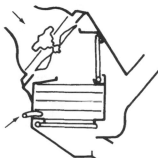

The drawings below show the cool air entries to the heater system. Using arrows show the flow of cool/heated air through each vehicle.

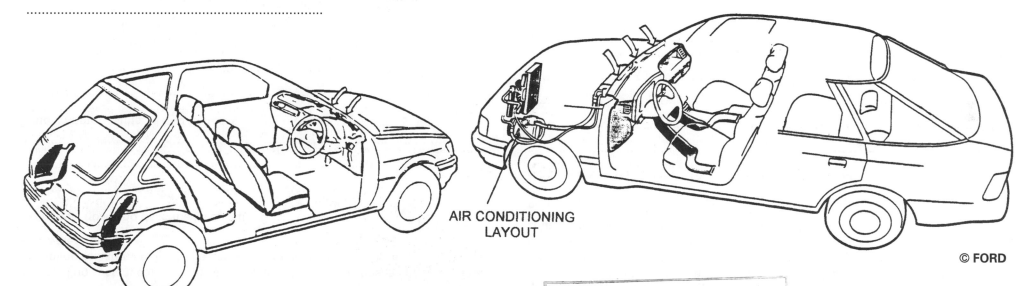

AIR CONDITIONING
LAYOUT

© FORD

HOSES

The cooling system components are connected by flexible hoses, that is, engine to cooling radiator, engine to heater radiator and possibly between cylinder head and water pump.

Examine a hose and describe its construction:

....

....

....

State common faults related to hoses:

....

....

....

Heater systems often require long hoses and on buses large air ducting layouts are also necessary.

Describe the heating arrangement of the bus shown.

....

....

....

....

....

....

....

....

....

....

....

....

....

© LEYLAND

SHUTTER CONTROLS

In very cold conditions it is particularly important that coolant temperature is maintained within very close limits. A cold running engine would considerably increase rates of engine wear. When cold conditions exist it can be desirable to fit shutters in front of the radiator to control the air flow through it.

State THREE ways in which the shutters may be automatically operated:

....

....

....

....

....

....

....

....

....

....

....

....

State the type of shutter operation shown and name the major parts:

....

EFFECTS OF FREEZING

When water freezes it increases in volume and this can cause a cylinder block to crack. State at what temperature cooling water increases its volume.

....

....

....

....

ANTIFREEZE

Fortunately the freezing point of water can be lowered considerably by the addition of certain liquids. An example of such a substance is ethylene glycol.

The recommended percentage antifreeze mixture in the UK is:

....

....

The graph shows two of the most important properties in an ethylene glycol-based antifreeze:

....

....

What other properties must an antifreeze possess?

....

....

What are the disadvantages of using a methanol-based antifreeze?

....

....

% Ethylene glycol in coolant

Graph: Temperature °C (y-axis) vs % Ethylene glycol in coolant (x-axis, 0 to 50)

The proportion of an ethylene glycol-based antifreeze present in a cooling system can be determined by checking the specific gravity of the coolant and by reference to its temperature.

What instrument is used to measure coolant antifreeze content?

....

....

....

....

Describe how the strength of antifreeze may be measured:

....

....

....

....

....

....

....

....

....

....

....

....

....

Name the parts indicated on the antifreeze hydrometer shown below:

ANTIFREEZE

State THREE operational tasks required when changing antifreeze:

1. ....

2. ....

3. ....

EFFECTS OF CORROSION – ACIDS AND ALKALIS

What are the main causes of corrosion in the cooling system?

....

....

....

....

....

....

....

How does continual heating affect the mineral deposits?

....

....

....

How can these corrosive elements in the system be neutralised?

....

....

....

COOLING SYSTEM PROTECTION DURING USE

Describe how the cooling and heater systems should be protected during use or repair by avoiding or preventing the following hazards:

(a) Corrosion

....

....

(b) Freezing

....

....

(c) Excess pressure/vacuum

....

....

(d) Coolant contamination

....

....

(e) Coolant loss

....

....

State any special tools that are required for carrying out cooling system maintenance:

....

....

....

....

....

DIAGNOSTICS: COOLING/HEATER SYSTEM – SYMPTOMS, FAULTS AND CAUSES

State a likely cause for each symptom/system fault listed below. Each cause will suggest any corrective action required.

SYMPTOM	SYSTEM FAULT	LIKELY CAUSE
System boiling, steam issuing from under bonnet.	Overheating	....
Engine is misfiring when accelerating and heater output is cool.	Overcooling	....
Engine whines when revving and there is a rumbling sound when idling.	Noisy water pump	....
Pool of liquid under vehicle and/or radiator empty.	Coolant leakage	....
Coolant discoloured and possible leakage at hose joints.	Corrosion	....
Coolant discoloured and appears to be oily.	Contamination	....
Coolant level at header tank is correct, heater fan works but there is no warm air coming into car.	Heater not operating	....
AIR COOLING The engine overheats when idling in traffic.	Air-cooling system overheating	....
AIR COOLING In cold weather the engine runs cold when cruising.	Air-cooling system overcooling	....

DIAGNOSTICS: COOLING SYSTEM NOISES

With the engine running what cooling system faults may be indicated by the following noises?

NOISE	POSSIBLE FAULT
Screeching noise when revving engine	
Buzz or whistle near radiator	
Rattling noise near radiator	
Ringing or grinding noise at front of engine	
Gurgling from radiator	

ROUTINE MAINTENANCE

List the general rules/precautions to be observed when carrying out the following routine cooling system maintenance and running adjustments:

1. Dealing with hot systems ....

....

....

2. Checking when engine is running ....

....

....

3. Antifreeze is spilt on paintwork. What action should be taken?

....

....

.... **NOTE: Antifreeze can be toxic.**

State FOUR reasons for carrying out routine maintenance on the cooling/heater system:

1. ....

2. ....

3. ....

4. ....

....

List typical routine maintenance/adjustment checks:

1. ....

2. ....

3. ....

4. ....

5. ....

6. ....

7. ....

8. ....

9. ....

10. ....

11. ....

12. ....

13. ....

14. ....

....

NOTE:

In your workshop class you may be expected to describe in detail any of the above checks.

HEAT LOSS CALCULATIONS

Heat losses occur when heat transfers from hot substances to cold substances.

....

If a small, medium and large engine were all operating at the same running temperature, which, when switched off, would cool down the quickest? ....

Heat energy is measured in .... It can be directly converted or expressed as mechanical work units where:

1 Newton metre = ...

The amount of heat contained in 1 kg of substance is known as its

....

The heat required to raise 1 kg of water 1 °C is ....

∴ The specific heat capacity of water is ....

Different materials accept (or lose) heat at different rates and therefore for a similar mass they will increase (or decrease) their temperature at different rates.

When heated at the same rate 1 kg of oil will increase its temperature at a

....rate than 1 kg of water.

Substance	Specific heat capacity (kJ/kg °C)	Substance	Specific heat capacity (kJ/kg °C)
Water		Steel	
Lubrication oil		Brass	
Aluminium		Lead	

List below the information it would be necessary to know, to be able to calculate the quantity of heat transferred from one substance to another:

....

Problems   **Note!** 1 litre of water has a mass of ....

Heat lost or gained/s	=	mass flow/s	×	specific heat capacity of substance	×	temperature change
Q	=	m	×	c	×	Δt

1. A pump circulates 150 litres of water through a cooling system in 2 minutes. The temperature at the top of the radiator is 90°C and at the bottom 70°C. Calculate the heat energy radiated per second.

....
....
....
....
....
....
....

2. An impeller unit circulates 2 litres of coolant to the radiator per second. Calculate the heat lost to air per second, when the temperature difference between top and bottom tank is 25°C.

....
....
....
....

3. (a) A cooling system contains 15 kg of water. Calculate the quantity of heat gained by the water if its temperature rises from 12°C to 88°C on starting.

....
....
....
....
....
....

3. (b) What heat is lost per second during cooling if the flow rate is 2 litres and the temperature at the bottom of the radiator is 54°C?

....
....
....
....

Chapter 4

Spark-Ignition Systems

Spark-ignition systems	51	Magnetic pickup or triggering system	59
Contact breaker coil-ignition system	51	Amplifiers – constant energy	59
Ignition coil	52	System checks – 'Lucas' constant energy	60
Ballast resistor	52	'Hall' effect system	61
Coil output	53	System check – 'Hall' effect	61
Primary circuit connections	53	Safety checks – electronic ignition	62
Mutual induction	53	Advantages of electronic systems	62
Distributor units	54	High-tension circuit	62
Contact breaker points	55	High-tension leads	63
Dwell angle	55	Suppressors	63
Capacitors	56	Spark plugs	64
Advance and retard mechanisms	56	Engine running conditions	65
Centrifugal advance	56	Routine maintenance	66
Vacuum advance	57	Diagnostics	67
Fault diagnosis: ignition system	58	Ignition-system protection during use	68
Electronic ignition systems	59		

SPARK-IGNITION SYSTEMS

Ignition systems with camshaft driven distributors may be conveniently divided into TWO types. These are:

1. ....

2. ....

One system, having contact breaker points and used on approximately pre-1980 vehicles, may be as shown opposite.

The other system may use one of many methods of allowing transistors to switch the primary circuit on and off (this being the function of the contact breaker points in the first system).

The most widely used methods of electronic ignition switching are:

1. ....   2. ....

Two other methods of switching are:

3. OPTICAL   and   4. CAPACITOR DISCHARGE

State the functional requirements of a coil-ignition system:

1. ....
....

2. ....
....

The coil-ignition system is basically divided into two circuits, PRIMARY and SECONDARY (low and high tension).

State the basic function of the main components shown opposite:

....
....
....
....
....
....

CONTACT BREAKER COIL-IGNITION SYSTEM

Complete the semi-pictorial wiring layout of a four-cylinder, coil-ignition circuit – shown below.

State a suitable firing order, ensure that the secondary circuit wiring complies with this firing order and draw a rotor in the distributor cap firing on number ONE cylinder.

Name the various parts.   Firing order ....

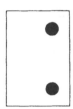

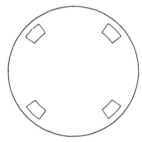

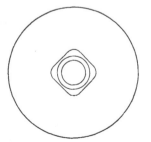

IGNITION COIL

Examine sectioned or dismantled ignition coils (that is, air-cooled, oil-filled and ballast-resistor types) then explain the operation of the coil shown below.

....

....

....

....

....

Name the parts in the sectioned sketch below which shows the basic coil wiring arrangement.

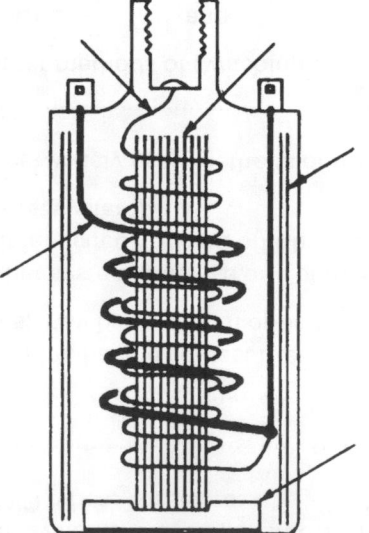

BALLAST RESISTOR

Most cars using contact breaker points incorporate a special low-voltage coil which during normal running is connected in series to a ballast resistor.

This ballast resistor may be placed in various positions in the ignition's primary circuit. Describe three resistors and their position in the circuit.

....

....

....

....

....

....

....

What is the reason for using a ballast resistor in the primary circuit?

....

....

Draw a primary circuit diagram to show the positioning of a ballast resistor.

To starter

Expain how such a low-voltage coil and ballast resistor operate.

....

....

....

....

....

....

....

What effect would leaving the ignition switched on (without the engine running) have on the ballast resistor?

....

Newer designs incorporate a relay cutout to prevent this problem.

COIL OUTPUT

The coil's secondary voltage builds up until it creates a spark at the plugs. When these are in good condition the voltage needed may be as low as ....
but with constant use and engine faults can rise to ....
The coil usually has a maximum output capacity of ....
With some transistorised coils it can be as high as ....

State what factors influence the ignition coil's output:

....

....

....

PRIMARY CIRCUIT CONNECTIONS

Most coils are wound and connected to produce a negative spark at the spark plugs.

Why is this? ....

....

....

....

....

....

Show the coil top markings when the primary circuit is wired correctly.

The effects of wiring the coil to give incorrect polarity would be:

....

....

....

....

....

....

....

....

MUTUAL INDUCTION

A simple rig is shown below. It demonstrates how an emf (voltage) can be induced into a circuit by varying the current flow in a neighbouring but separate circuit. Describe how to induce and vary an emf.

....

....

....

Name the parts.

Describe how an emf is induced by mutual induction:

....

....

....

....

....

Use an ignition coil in a similar manner to produce a spark.

Constructionally in what way does the coil differ from the apparatus above?

....

....

....

DISTRIBUTOR UNITS

The purpose of a distributor can be divided into three distinct roles. These are to:

1. ....
   ....
   ....
   ....
   ....

2. ....
   ....
   ....
   ....

3. ....
   ....
   ....
   ....

The method of interrupting the flow of primary current creates considerable variations in design, as can be seen on the distributors shown on this page. However, the methods of high-voltage distribution and the means of advancing the point of ignition remain basically the same on each distributor.

Name the main parts of the distributor units shown and identify the type of ignition system for which they would be suitable.

© **LUCAS**

© **FORD**

SYSTEM

CONTACT BREAKER POINTS

In relation to contact breaker points, at what instant in their operation is the spark at the plug produced?

....

Describe the operation of the contact breakers.

....

....

....

....

....

....

The contact breaker points are positioned on the base plate of the distributor. The movable point is operated by the cam. Identify these items on the drawings below and state typical gap settings.

Make ....

© VAUXHALL

Recommended point gap setting ....

© LUCAS

Recommended point gap setting ....

DWELL ANGLE

What is meant by 'dwell angle'?

....

....

....

....

....

....

Complete the drawing to show dwell angle.

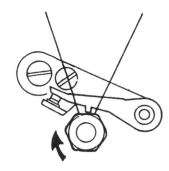

What effect would the following incorrect dwell angle have on coil performance and therefore the running of the engine?

Excess dwell: ....

....

Insufficient dwell: ....

....

Quote different manufacturers' specifications for four-cylinder distributors fitted to various vehicles.

Distributor make	Vehicle		Dwell angle	CB gap setting
	make	model		

Replace points and check dwell angle/CB gap setting on various distributors.

CAPACITORS

What is the function of the capacitor in the contact breaker system?

....
....
....

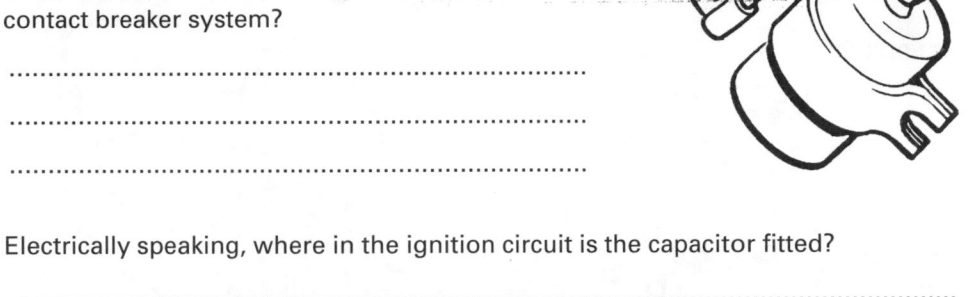

Electrically speaking, where in the ignition circuit is the capacitor fitted?

....
....
....

The capacitor consists of two sheets of tin foil or metallised paper which are rolled together but electrically insulated from one another.

Name the parts on the sketches shown below:

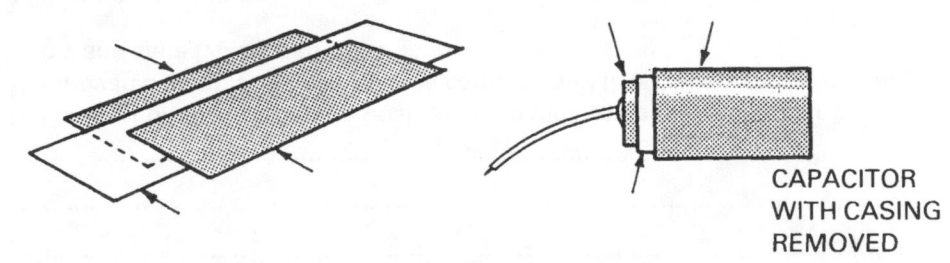

CAPACITOR
WITH CASING
REMOVED

State three capacitor faults and their symptoms, which may cause ignition troubles.

1. Fault ....

   Symptom ....

2. Fault ....

   Symptom ....

3. Fault ....

   Symptom ....

ADVANCE AND RETARD MECHANISMS

© **LUCAS**

When the engine speed or load is varied it is usual to adjust the ignition timing automatically. This may be done in two ways:

1. ....
....
....

2. ....
....
....

CENTRIFUGAL ADVANCE

The type shown above has pivoted weights mounted below the contact breaker base plate.

Examine similar distributors and explain their principle of operation.

....
....
....
....
....
....
....
....

VACUUM ADVANCE

Complete the drawings to show the operation of the vacuum unit and its connection to the distributor base plate and carburettor.

Name the various parts.

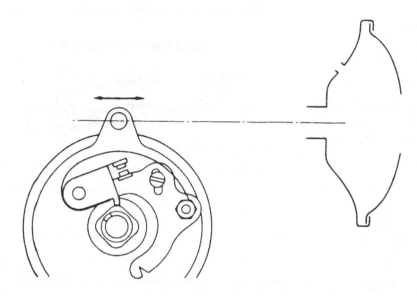

Explain how this type of advance and retard mechanism operates.

....

....

....

....

....

....

TWIN DIAPHRAGM VACUUM UNIT FOR EMISSION CONTROL

This type uses a diaphragm to advance the ignition timing in the conventional way and has a second diaphragm to positively retard the timing. The vacuum connection for this second diaphragm is positioned in the inlet manifold downstream from the throttle.

What is the function of this unit?

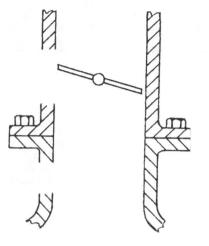

....

....

....

....

....

SPARK SUSTAIN AND SPARK DELAY VALVES

The diagram shows the vacuum pipe layout between distributor and carburettor (Ford). It incorporates valves used to improve emission control by modifying the vacuum advance action in the distributor. Name the arrowed components.

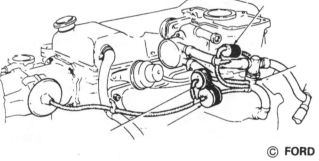

© FORD

What is the function of the:

Spark Delay Valve ....

....

Spark Sustain Valve ....

....

FAULT DIAGNOSIS: IGNITION SYSTEM – ENGINE WILL NOT START

Contact Breaker Point Systems

Preliminary inspection:

1. Check all connections including battery.
2. Check battery state of charge.
3. Check plugs, HT leads and cap condition.

How would you test initially for an HT spark?

....

....

....

....

....

....

....

....

....

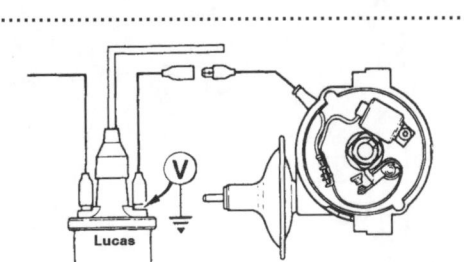

SPARK PLUG SIMULATOR

Carry out a Primary Circuit Test in the workshop on an available vehicle using specific manufacturer's instructions.

The following diagrams indicate the general tests that all primary circuits require.

State what each diagram is indicating.

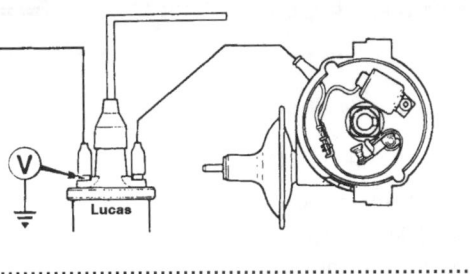

Test 1 ....

....

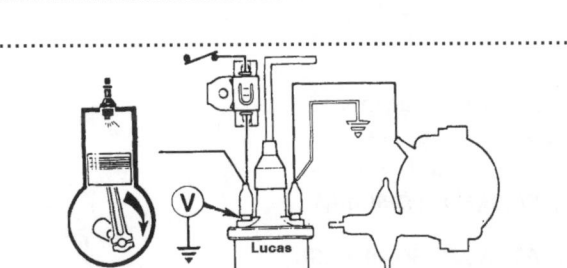

Test 2 ....

....

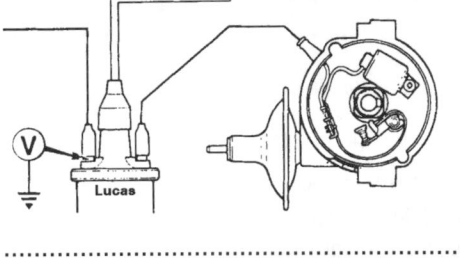

Test 3 ....

....

....

....

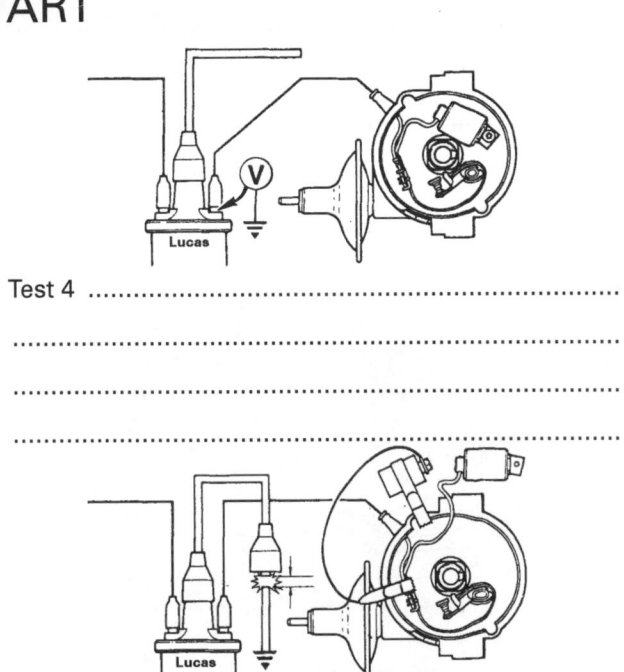

Test 4 ....

....

....

Test 5 ....

....

....

Test 6 ....

....

....

ELECTRONIC IGNITION SYSTEMS

Electronic ignition systems use power transistors as high-speed switches in the circuit, carrying the heavy current normally handled by the contact breaker points.

With the contactless systems the contact breaker is removed, and replaced with a trigger system, known as a trigger wheel, timing rotor or reluctor. The only moving parts in the distributor that may then wear are the advance–retard mechanisms.

MAGNETIC PICKUP OR TRIGGERING SYSTEM

The sketches show two types of inductive coil or permanent magnet triggering designs. Name the operating parts and describe their switching action.

'Limb' magnetic coil   Annular magnetic coil

'LIMB' MAGNETIC COIL

....

....

....

....

ANNULAR MAGNETIC COIL

....

....

....

....

The diagram shows the wiring layout of an Austin Rover Metro 1.3 constant energy, annular magnetic coil type electronic ignition system.

Name the basic parts and complete the wiring colour grid.

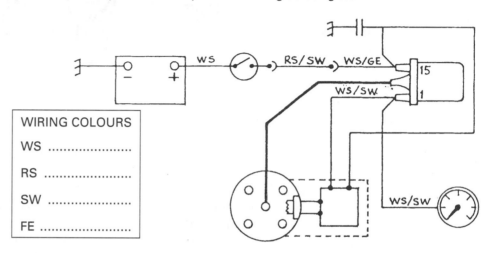

WIRING COLOURS	
WS	....
RS	....
SW	....
FE	....

AMPLIFIERS – CONSTANT ENERGY

The designs, which are constantly improving, can be divided into THREE types: constant dwell, constant energy and mapped or programmed types. The ones shown opposite are of constant energy design.

What is the purpose of the amplifier?

....

....

....

....

SYSTEM CHECKS – 'LUCAS' CONSTANT ENERGY

Preliminary inspection:

1. Check all connections including battery.
2. Check battery state of charge.
3. Check LT and HT external circuits (see also Test 7).
4. Check pickup air gap setting.

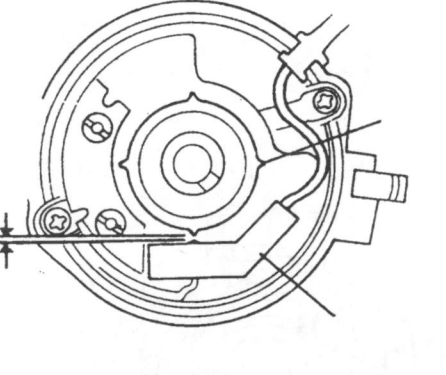

LIMB TYPE MAGNETIC PICKUP

State an important point when checking this air gap:

....

....

....

Test 1. HT SPARKING TEST
Carry out HT Sparking Test as described on page 58.

If good spark occurs go to Test 6, page 58.

Describe the checks shown in diagrams.

Test 2. AMPLIFIER STATIC CHECK

....

....

....

....

Test 2

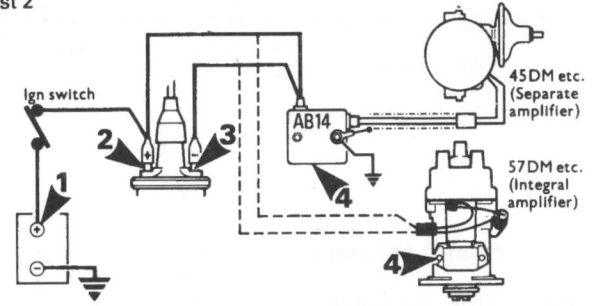

45DM etc. (Separate amplifier)

57DM etc. (Integral amplifier)

EXPECTED READINGS

1　More than 11.5 volts
2　1 volt max below volts at 1
3　1 volt max below volts at 1
4　0–0.1 volt

1	2	3	4	SUSPECT
L	✓	✓	✓	Discharged battery
✓	L	L	✓	Ign. switch and/or wiring
✓	✓	L	✓	Coil of amplifier
✓	✓	✓	H	Amplifier earth

Test 3. CHECK AMPLIFIER SWITCHING

....

....

....

....

Test 3

To ign switch

To distributor

Crank engine

12V

Test 4. CHECK PICKUP COIL RESISTANCE

Expected resistances
Lucas 2000–5000 Ω pre 86
　　　950–1100 Ω 86 on
Bosch　630–1100 Ω

Test 4　　　　　　　　　　　　© **LUCAS**

Separate amplifier　　In-built amplifier

45DM　　　　　　　　　　Disconnect　　57DM

....

....

....

....

If engine will still not start.
Check HT spark (Test 5, page 58) again.
If no spark replace coil.
If good spark then check rotor arm (Test 6, page 58).

....

....

Test 7. VISUAL AND HT CABLE CHECK

Examine:　　　1. Distributor cover
　　　　　　　2. Coil top
　　　　　　　3. HT cable insulation
　　　　　　　4. HT cable continuity
　　　　　　　5. Sparking plugs.

SAFETY
Some amplifier units contain Beryllia and should not be crushed or opened in any way.

'HALL' EFFECT SYSTEM

The 'Hall' effect generator has a trigger which incorporates a semiconductor material enclosed by a permanent magnet. When a voltage passes through the semiconductor it generates an independent small current at right angles to the magnetic field. This is supplied to the amplifier.

What then occurs as the drive shaft turns?

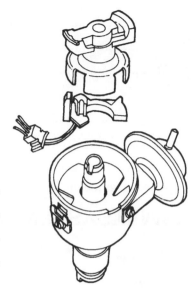

....

....

....

The diagram below shows the wiring layout of a Ford Fiesta 1.3 Programmed 'Hall' effect type electronic ignition system.

Name the basic parts and complete the wiring colour grid.

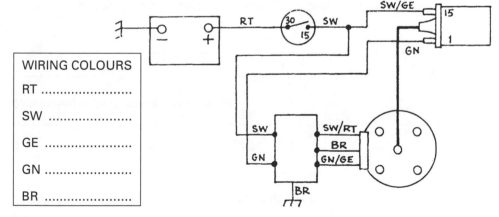

WIRING COLOURS
RT ....
SW ....
GE ....
GN ....
BR ....

NOTE

The wiring diagram is schematic. The amplifier is mounted on the bulkhead. The Hall effect pickup in the distributor is not replaceable; if faulty, a new distributor is required.

SYSTEM CHECK – 'HALL' EFFECT

Preliminary checks same as constant energy type (no air gap check).

Test 1. HT SPARKING TEST
      If strong spark go to Test 2.
      If no spark go to Test 3.

Test 2. VISUAL AND HT CABLE CHECK
      (see Test 7, page 60)

Test 3. CHECK CONDITION AND CONTINUITY
      OF WIRING BETWEEN BATTERY AND COIL

Test 3

Test 4

....

....

....

....

....

Test 4. CHECK OPERATION OF AMPLIFIER AND PICKUP

....

....

....

....

....

....

Test 5. REPEAT Test 4, having obtained a new distributor.
      Spin distributor, bulb should now flash.

Test 6. If bulb still does not flash and all tests have been satisfactory, fit new amplifier.

NOTE This is a very brief and general description.
When carrying out such tests in the workshop, more detailed manufacturer's instructions for the vehicle concerned must always be used.

SAFETY CHECKS – ELECTRONIC IGNITION

In order to protect yourself or the system what should be done when:
1. disconnecting and reconnecting cables?

....

2. holding removed spark plug leads with engine running?

....

....

....

3. working on Hall effect distributors?

....

....

....

When fitting amplifiers what important earthing procedure should be observed?

....

....

....

ADVANTAGES OF ELECTRONIC SYSTEMS

State the advantages of electronic ignition when compared to contact breaker types:

....

....

....

....

....

HIGH-TENSION CIRCUIT

The high-tension circuit consists of the coil, distributor cap, rotor, plugs and plug leads. Describe the flow of the high voltage current.

....

....

....

....

DISTRIBUTOR CAP

Two ways in which the high-tension lead may be held in the distributor cap are:

(i) .... (ii) ....

State what material the distributor cap is made from and give reasons for its use:

....

....

....

What electrical property does the rotor contact brush that is fitted into the cap often incorporate?

....

ROTOR

On certain high-speed engines the maximum speed is controlled by a rotor cut-out. How does this operate?

....

....

....

....

HIGH-TENSION LEADS

Three types of leads are commonly used:

1. The centre core is made of stranded copper wire.

2. The centre core is made of stranded and woven rayon or silk which is impregnated with graphite.

3. The centre core is made of current-conducting glass fibre surrounded by extra insulating material.

What are their relative advantages?

1. Copper core leads ....

....

2. Graphite carbon core leads ....

....

3. Reactive cable ....

....

Explain how the terminal connections are made on the graphite-impregnated leads:

....

....

....

Name the two types of high-tension coil connections shown:

....

SUPPRESSORS

Whenever an electrical spark occurs, waves of electrical energy are radiated out from the spark source. This energy causes external electrical interference which by law must be suppressed (Wireless Telegraphy – Control of Interference from Ignition Apparatus – Regulations 1952).

Resistors having a value of between 5000 and 25,000 ohms are required; these may be incorporated in at least four components.

1. ....

2. ....

3. ....

4. ....

State the statutory requirements relating to installation and functioning of the ignition system in terms of radio interference suppression.

....

....

....

Explain the effect 'flashover' (spark tracking) has on the high-tension circuit.

....

....

....

....

....

....

SPARK PLUGS

On the sectioned view of the spark plug shown, the leader lines indicate certain special features. Comment on these features.

Insulator ....

....

....

....

....

....

Centre electrode ....

....

....

....

....

Sillment seal ....

....

....

Gasket ....

....

Earth electrode ....

....

© CHAMPION

What are the effects of running an engine with the plug electrodes set:

1. too wide? ....

....

2. too close? ....

....

The high-voltage energy produced by the coil is dissipated in the form of a spark across the spark plug electrodes. This ignites the petrol/air mixture in the combustion chamber.

The spark plug must operate efficiently under widely varying conditions of pressure and temperature and must be designed to suit the type of engine to which it is fitted.

State the importance of the following variations in plug design:

Plug reach – long or short ....

....

....

Thread diameter ....

....

....

Methods of sealing plug seat ....

....

....

....

Installation of:

Both designs of plug should be tightened down by hand until finger tight, and then tightened by a plug spanner by a further amount:

1. Gasket type

....

2. Taper seat

....

1. Gasket type
2. Taper seat type

Indicate on plan view the amount to be tightened.

Variation in Plug Design

HEAT RANGE

For any particular application, spark plugs must be selected that do not foul at slow speeds or get red hot at high speeds.

Describe how heat range is designed into a spark plug.

....
....
....
....
....
....
....

State which plug below is the hot and cold plug, and indicate with arrows the heat flow path on both plugs.

....

ENGINE RUNNING CONDITIONS

If possible, examine a manufacturer's spark plug condition colour chart, together with uncleaned spark plugs removed from various engines.

State the plug's likely appearance when the engine's running condition is said to be:

Normal	....
Mixture too weak	....
Mixture too rich	....
Engine overheated	....

Examine a manufacturer's chart giving spark plug specifications.

Select a range of plugs and complete the table below.

Make of spark plugs considered ....

Plug number, place coldest plug first	A typical vehicle for which plug is recommended	Correct gap setting

Examine the chosen spark plug chart and note what each number and letter signify.

ROUTINE MAINTENANCE

State reasons for carrying out routine maintenance on the ignition system:

1. ....

2. ....

3. ....

4. ....

List TEN routine maintenance/service checks:

1. ....

2. ....

3. ....

4. ....

5. ....

6. ....

7. ....

8. ....

9. ....

10. ....

(Note: 'check' means also 'adjustment or renew if necessary'.)

Inspections

1. In the ignition circuit the most common items associated with electrical resistance are the ignition leads and ignition coil. Using an ohm-meter (multi-meter), measure the resistance values of such items.

	King	1	2	3	4	Recommended values
Ignition leads						
Coil Primary	coil one		coil two			1 .... 2 ....
Coil Secondary						

2. Describe a method of checking the ignition timing using a stroboscopic device.

....

....

....

....

....

....

....

....

....

3. Describe the method of statically checking the ignition timing, using a test lamp.

....

....

....

....

4. Check the operation of mechanical and vacuum advance mechanisms.

(a) Accelerate engine and note maximum amount of advance. This is both the mechanical and vacuum advance.

Actual reading .... Manufacturer's spec. ....

(b) Disconnect the vacuum pipe and accelerate engine. This will give the mechanical advance only.

Actual reading .... Manufacturer's spec. ....

DIAGNOSTICS: SPARK IGNITION SYSTEM – SYMPTOMS, FAULTS AND CAUSES

State a likely cause for each symptom/system fault listed below. Each cause will suggest any corrective action required.

SYMPTOM/SYSTEM FAULTS	LIKELY CAUSE
Engine is poor at starting, it cranks but does not fire for a long time.	....
Engine misfires and runs rough when accelerating.	....
Engine suddenly cuts out, particularly when vehicle is running on a bumpy road.	....
Engine misfires and hesitates when revving.	....
There is a high pitched tapping or rattling noise from the engine when the vehicle is accelerated.	....
More petrol than usual is being used since the last service.	....
Vehicle seems very slow and sluggish and low on power.	....
Engine is running very hot.	....
Engine continues to run when switched off.	....
Radio crackles when the engine is running.	....

IGNITION-SYSTEM PROTECTION DURING USE

Describe how the ignition system should be protected during use or repair from the following hazards:

Dirt and moisture ....

....

Avoiding neglect ....

....

The semiconductors and transistors in an electronic ignition system are highly sensitive to incorrect connections and voltage surges.

What precautions should be taken relative to the following?

Battery connections ....

....

....

....

....

Making connections to electronic systems ....

....

....

....

....

....

Other work on the vehicle ....

....

....

Describe any special tools required to carry out routine maintenance and running adjustments:

....

....

....

....

....

....

....

....

....

....

....

List the general rules/precautions to be observed while carrying out routine ignition-system maintenance and running adjustments when working:

1. on the ignition system with the engine running

....

....

....

2. near rotating parts ....

....

....

3. in an enclosed space with the engine running

....

....

Chapter 5

Air Supply and Exhaust Systems

Pollutants created by the motor vehicle 70
Air cleaner-silencer basic types 70
Thermostatically controlled air inlet valves 72
Evaporative emission control 73
Air supply – protection during use 73
Manifolds 74
Hot spot control 74
Electric manifold heating 74
Manifolds – methods of improving fuel economy,
   performance and emissions 75
Exhaust systems 76

Flexible joints 77
Catalytic converter 78
Lambda (oxygen) sensor 79
Exhaust gas recirculation (EGR) 79
Secondary air injection system 79
Exhaust emissions 80
Pollution effects on the environment 80
Exhaust system – protection during use 81
Routine maintenance 81
Statutory requirements 81

POLLUTANTS CREATED BY THE MOTOR VEHICLE

The creation of some pollution is inevitable when power is obtained by internal combustion using a hydrocarbon fuel; and when an engine is not running in tune, an excess of harmful emissions is passed into the atmosphere.

Name the three basic areas that emit pollutants into the atmosphere:

....

....

....

Name the pollutants produced by an engine and state why they are there:

....

....

....

....

....

....

....

....

AIR CLEANER–SILENCER BASIC TYPES

Most modern air cleaner–silencers are of the paper element type. They are usually mounted on top of the engine intake (Carburettor models) or fixed to the side of the engine compartment (Fuel injection models).

Name four types of air cleaner:

1. .... 3. ....

2. .... 4. ....

Four requirements of the assembly are:

1. ....

....

2. ....

3. ....

4. ....

Paper Element Type

Two examples of dry paper element filters are shown below.

Name the parts indicated:

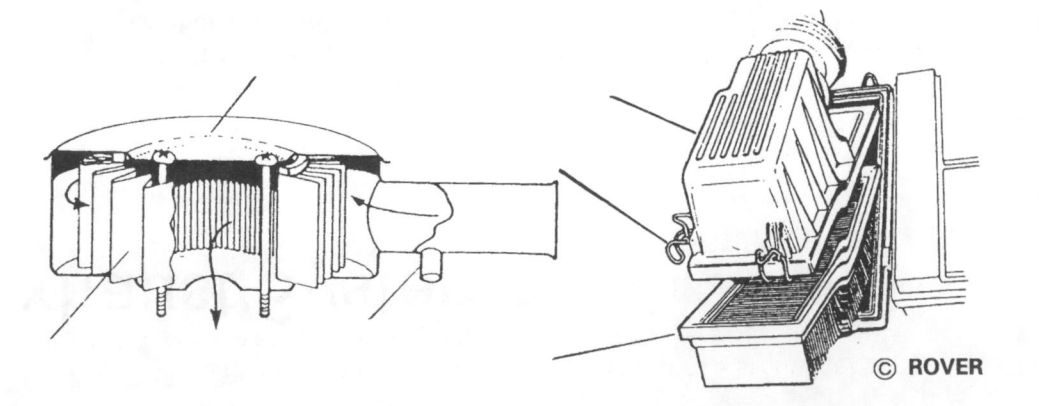

© ROVER

Oil Bath Air Cleaner

An oil-bath-type air cleaner is shown below in section. Note the oil in the base and the oil-wetted wire mesh filter. Indicate the air flow through the unit and name the main parts.

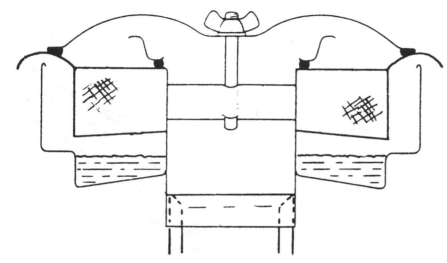

Describe the servicing procedure for the oil filter above:

....

....

....

....

....

What would be the effects of a partially blocked air cleaner?

....

Centrifugal Pre-cleaner

In very dusty conditions a heavy duty DRY air cleaner with a secondary paper element filter such as shown (sectioned) below may be used.

Indicate the air flow through the cleaner, name the parts and state briefly how it works.

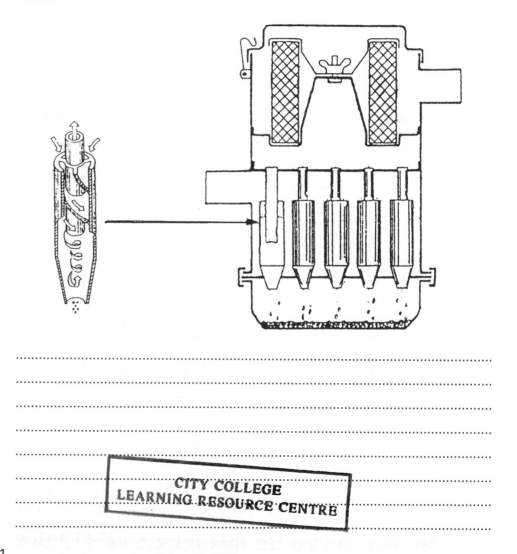

....

....

....

....

....

....

....

THERMOSTATICALLY CONTROLLED AIR INLET VALVES

Modern vehicles are designed to have the air warmed before it enters a cold engine. This is achieved by having the air pass over the surface of the exhaust manifold before it enters the intake ducting. When the engine is warm cooler air is preferred and so a flap valve shuts off the supply of hot air.

Why is it desirable to control the air intake temperature?

....

....

....

....

....

A temperature-sensitive flap valve is shown below. Name the main parts and describe the valve's operation.

© **ROVER**

....

....

....

....

....

....

....

An alternative method of opening is to have a flap valve operated by a waxstat (wax operated thermostat). How would this system operate?

....

....

....

A vacuum operated air control flap valve is shown below. Name the main parts and describe the valve's control action when the engine is cold and at its normal operating temperature.

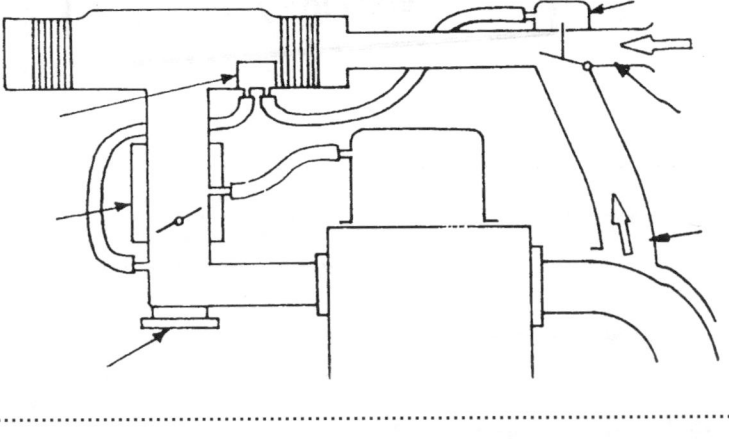

....

....

....

....

....

....

....

Describe how the operation of the air temperature control valve can be checked:

....

....

....

....

....

....

EVAPORATIVE EMISSION CONTROL

Having recognised the need to control pollution, it is relatively easy to control crankcase emissions and fuel loss through evaporation.

How are these emissions controlled? ....

....

....

Identify the items indicated:

© ROVER

CHARCOAL CANISTER

A charcoal canister filter is positioned in the vapour line system when a greater control of fuel evaporation is required. How does the canister operate?

....

....

....

FOR CRANKCASE EMISSION CONTROL SEE PAGE 27

AIR SUPPLY – PROTECTION DURING USE

Describe how the air supply system should be protected during use or repair from the following hazards:

(a) Contamination of filter by crankcase emissions

....

....

....

....

(b) Exceptionally dusty conditions

....

....

State reasons for carrying out routine maintenance on air supply systems:

1. ....

....

2. ....

....

List typical routine maintenance/adjustment checks:

1. ....

2. ....

3. ....

4. ....

5. ....

6. ....

....

MANIFOLDS

Inlet manifolds allow the air/fuel mixture to be distributed from a single duct to the many required branches, that is, 4-cyl., 6-cyl., 8-cyl. etc. engines. The exhaust manifold provides the flow in the reverse manner.

A 4-cylinder fuel injection engine manifold layout is shown below. Identify the items indicated.

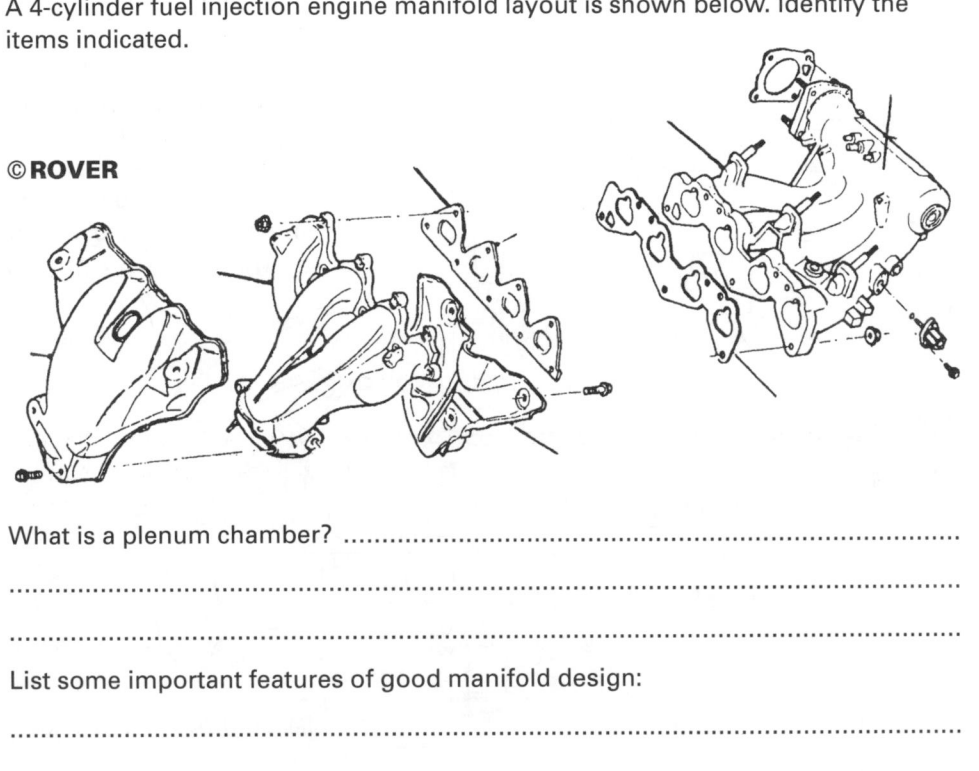

© ROVER

What is a plenum chamber? ....

....

....

List some important features of good manifold design:

....

....

....

....

....

Why is some form of heating of the inlet manifold considered necessary?

....

....

HOT-SPOT CONTROL

The sketch shows a thermostatically controlled hot-spot device. This allows for a rapid warm-up and then ensures the exhaust gas by-passes the hot spot.

Indicate which drawing shows the hot-spot control valve in its open position and closed position. Name the parts.

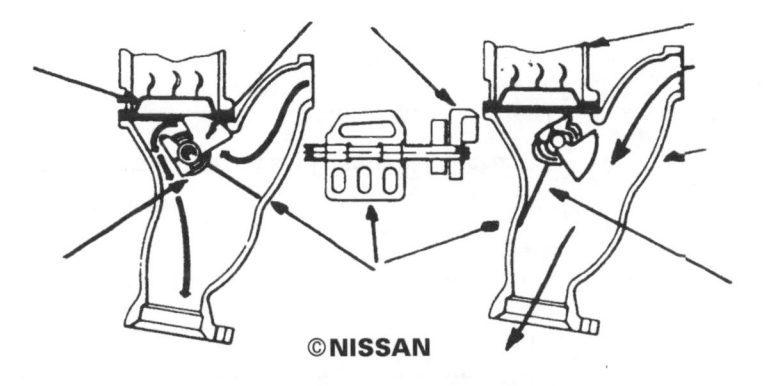

© NISSAN

ELECTRIC MANIFOLD HEATING

Manifold heating which is instant when a cold engine is started is by an electrical resistance. A common example found on modern vehicles is a circular pad fitted at the underside centre of the inlet manifold.

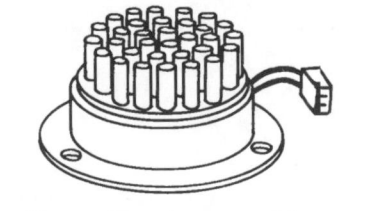

Flat type used in: Pin cushion type used in:

....

Why is the current for these heaters supplied from the engine oil pressure relay?

....

....

MANIFOLDS – METHODS OF IMPROVING FUEL ECONOMY, PERFORMANCE AND EMISSIONS

High performance engines using four valves per cylinder may improve fuel economy and performance by controlling the air input into the cylinder in different ways. Three examples are:

1 By preventing one of the two inlet valves from opening at low speeds. This can be achieved by using oil pressure controlled spool valves in the rocker assembly to lock together the rockers (one free wheeling and the other operating) of each pair of valves, so that at a predetermined speed and load they both operate.

(The mechanism is similar to the HONDA variable valve timing, page 179.)

2 Changing the effective length of the induction tract between low and high speed.

3 Restricting the air flow to one inlet valve. This has a similar effect to 2.

Gas flow into the engine should be timed so that it is moving forward when the inlet valve is open and so can ram into the cylinder. The varying engine speed, pumping action of the piston and closing of the valves all act against this ram effect happening easily. The air movement inside the intake passages bounces and creates pressure waves which vary greatly between low and high speeds. All three examples above are ways of taking advantage of the air's natural movement.

CHANGING LENGTH OF INDUCTION TRACT

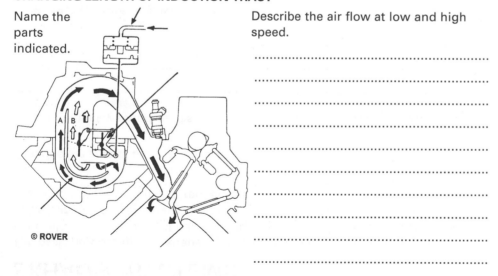

Name the parts indicated.

© ROVER

Describe the air flow at low and high speed.

....
....
....
....
....
....
....
....
....
....

RESTRICTING THE AIR FLOW TO ONE INLET VALVE

Name the parts indicated.

© PEUGEOT

Describe with the aid of the numbers on the diagram the air flow at low and high speed.

....
....
....
....
....
....
....
....
....

What effects do all these designs have on engine performance?

....
....

DOUBLE SKINNED EXHAUST MANIFOLD PIPES

To meet EU 2000 Emission Standards double skinned exhaust manifold pipes are used. Why has this arrangement become necessary?

© HONDA

....
....
....
....

EXHAUST SYSTEMS

The basic purpose of the exhaust system is to silence the noise created by the high velocity of the exhaust gas (around 100 m/s) as it leaves the engine. The noise is in the form of varying frequencies, and different types of boxes are designed to cope with these frequencies.

Describe the type of box that absorbs:

1. low-frequency sound waves ....

....

....

....

2. high-frequency sound waves ....

....

....

....

Identify the types of box shown.

....

....

....

Indicate the gas flow through the first box.

Which noise frequencies cause the most concern? ....

....

The complete system must be flexibly mounted on the chassis; and at the attachment to the exhaust manifold there must be a gas-tight seal.

Name the parts indicated on the exhaust system shown below.

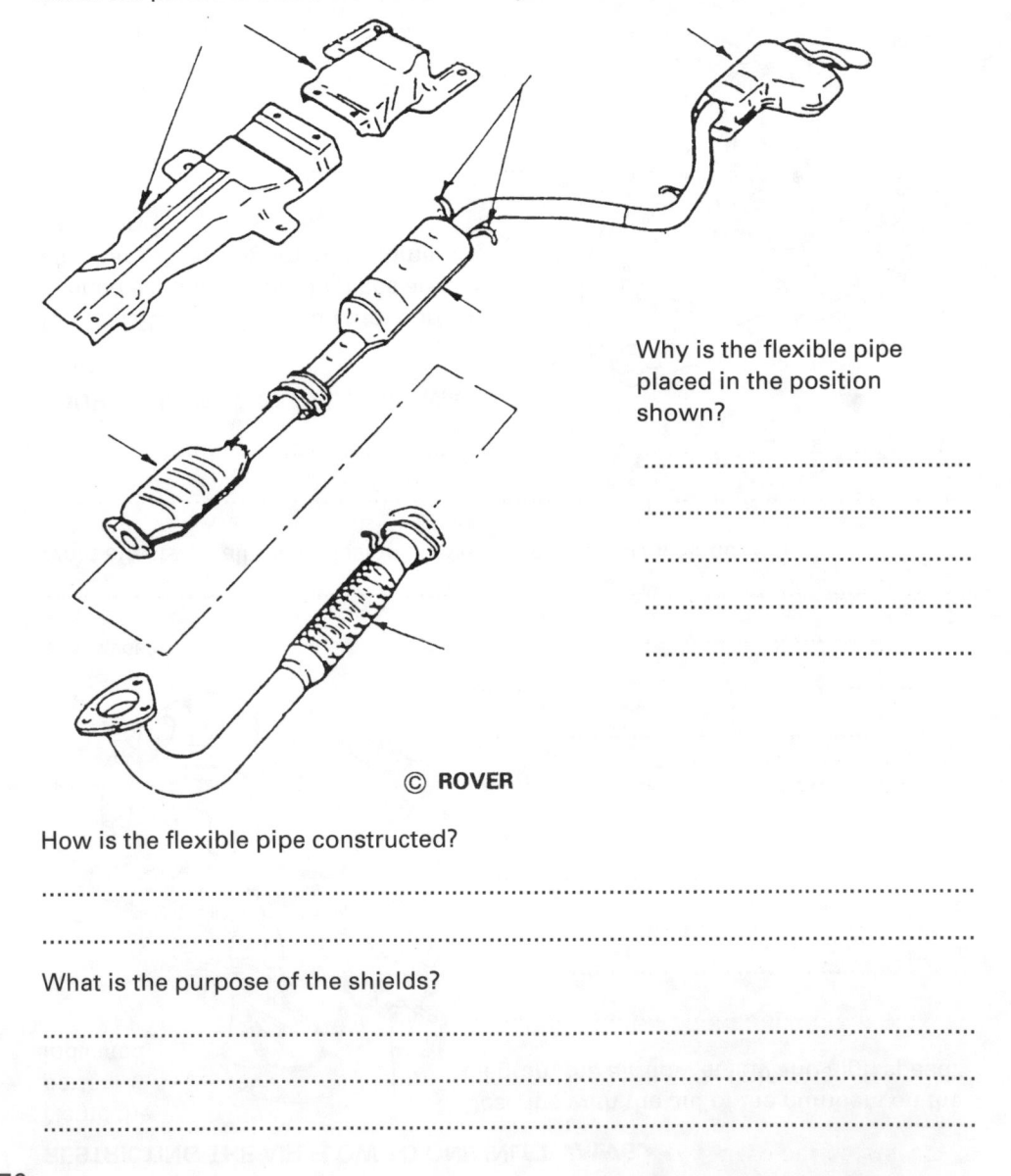

© ROVER

Why is the flexible pipe placed in the position shown?

....

....

....

....

....

How is the flexible pipe constructed?

....

....

What is the purpose of the shields?

....

....

FLEXIBLE JOINTS

Flexible joints and pipes are special features which may be used in exhaust systems when vibration or engine rocking is a problem.

Describe the flexible layout shown.

....

....

....

....

....

....

....

....

....

....

....

Advantages ....

....

Disadvantages ....

An exhaust system corrodes on the outside. Why does it also corrode from the inside outwards?

....

....

....

How can internal corrosion be reduced?

....

....

The part sectioned flexible connection shown below has a loose thick metal collar fitted in the joint; it is secured by bolts, coil springs and nuts.

Why should the fixed spring length be (in this case) 25 mm as shown?

METAL COLLAR

25

....

....

....

....

....

....

....

....

....

....

....

....

....

....

....

State FOUR purposes or functional requirements of an exhaust system:

1. ....

....

2. ....

....

3. ....

....

4. ....

CATALYTIC CONVERTER

A catalytic converter provides a means of considerably reducing harmful exhaust emissions to below the legally required limit.

All new vehicles are required to fit catalytic converters.

Identify the catalytic converter and state the position of the lambda probe on the vehicle shown.

Corrugated metal foil type catalyst matrix

Its design is more compact than the ceramic type and is suitable for use with small engines.

Why must the converter be fitted near the front of the system?

....

....

Describe the contents of the converter and the chemical reaction that occurs.

....

....

....

....

....

Why must leaded fuel NEVER be used in engines using converters?

....

....

....

Sketches below show sections through a catalytic box. Name the main parts, indicate the gas flow path and state the harmless gases that leave the converter.

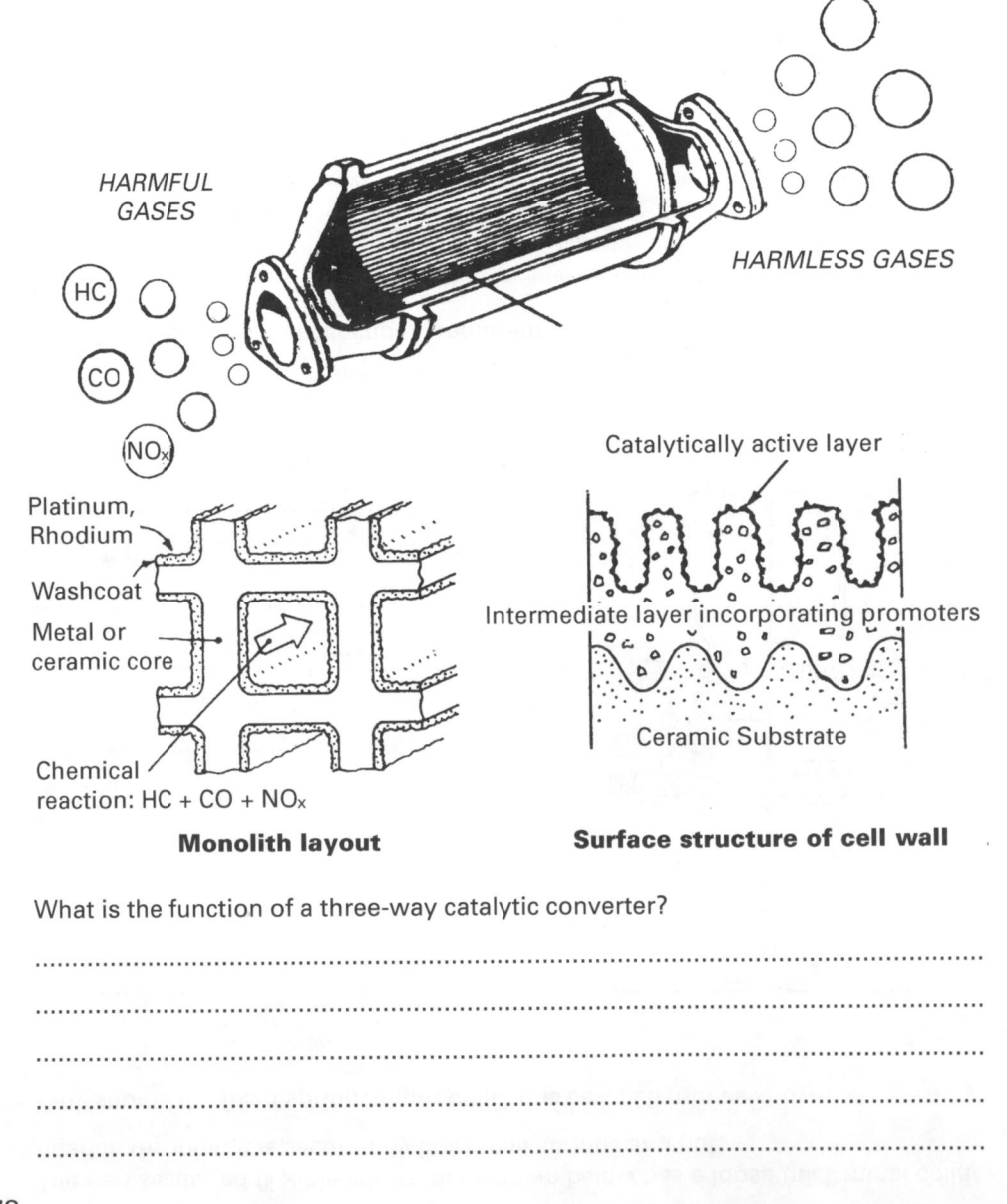

HARMFUL GASES

HC

CO

NOx

HARMLESS GASES

Catalytically active layer

Platinum, Rhodium

Washcoat

Metal or ceramic core

Intermediate layer incorporating promoters

Ceramic Substrate

Chemical reaction: HC + CO + NOx

Monolith layout

Surface structure of cell wall

What is the function of a three-way catalytic converter?

....

....

....

....

How is the catalytic converter affected by temperature?

....
....
....
....
....
....
....

What are the disadvantages of catalytic converters?

....
....
....
....

LAMBDA (OXYGEN) SENSOR

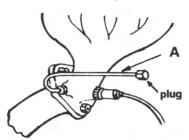

To ensure that the exhaust gas contents are kept within reasonable limits a Lambda sensor is fitted to the manifold where it joins the exhaust system.

To what would pipe 'A' be connected when the blanking plug was removed?

....

If the air fuel ratio entering the combustion chamber is too rich the exhaust gasses will be low in oxygen and the sensor will send a low voltage signal to the ECU, as the mixture weakens more oxygen becomes present and the voltage signal rises. The aim of the system is to maintain a chemically correct air/fuel ratio of 14.7:1 by weight (the stoichiometric ratio) at all times.

EXHAUST GAS RECIRCULATION (EGR)

This is an emission control commonly used to allow up to 15% of the exhaust gas to recirculate back into the inlet manifold. Why is this considered necessary?

....
....

Identify the EGR components on the layout below

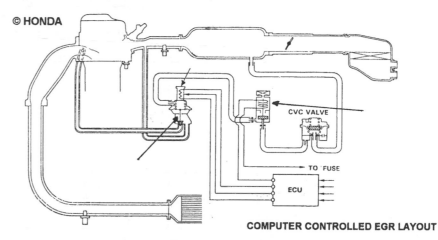

COMPUTER CONTROLLED EGR LAYOUT

The exhaust gas recirculation EGR system consists of an EGR diaphragm valve, connected by pipes between the exhaust and inlet manifold, and an EGR solenoid (modulator) valve and connecting vacuum hoses. The solenoid valve is controlled by the engine ECU which will operate the EGR valve and can vary the flow relative to engine speed, load and temperature requirements.
How is this flow control variation achieved?

....

(See also page 275.)

SECONDARY AIR INJECTION SYSTEM

To decrease exhaust emissions when the engine is cold and to ensure a quick warm up of the catalytic converter, air can be blown into the exhaust manifold at points near to each exhaust valve. This secondary air injection system consists of a pump, air cut-off valve and solenoid valve and exhaust non-return valve. The system is controlled by the engine's ECU. After warm up the pump switches off.
How does this air cause a rapid warm up during the engine's warm up cycle?

....
....
....

EXHAUST EMISSIONS

Exhaust gas is the product of the combustion of air and fuel mixture.

Air consists mainly of ....

Fuel consists mainly of ....

State how these products combine to form the exhaust gas.

....

....

....

....

State the harmless exhaust gas products:

1. ....   3. ....

2. ....   4. ....

List the main undesirable exhaust products and state the cause of each:

....

....

....

....

....

....

....

....

POLLUTION EFFECTS ON THE ENVIRONMENT

Rightly or wrongly, it is suggested that around 40% of all sources of air pollution are attributable to motor vehicles. They are said to produce roughly 50% man-made hydrocarbons, 60% of all carbon monoxide and 40% oxides of nitrogen.

Complete the chart to state the effects of the main pollutants produced by motor vehicles.

Pollutant	Symbol	Effects on people	Effects on the environment
Carbon monoxide		....	....
Carbon		....	....
Hydrocarbon		....	....
Lead		....	....
Oxides of nitrogen		....	....

EXHAUST SYSTEM – PROTECTION DURING USE

Describe how the exhaust system is protected from the following hazards during use or repair:

(a) Preventing corrosion ....
....
....

(b) Preventing external mechanical damage ....
....
....

(c) Avoiding overheating of other parts ....
....

(d) Preventing blowing of joints ....
....

(e) Types of fuel ....
....

ROUTINE MAINTENANCE

State reasons for carrying out routine maintenance on exhaust systems:

1. ....
2. ....
....
3. ....
4. ....
....

List typical routine maintenance visual checks:

1. ....
2. ....
3. ....

List some general rules/precautions to be observed while carrying out routine air–fuel/exhaust system maintenance when:

© BETEX

1. Running an engine in a confined space
....
....
....
....
....

2. Handling hot components ....
....

3. Working over unguarded air intakes ....
....

4. Completing the repair job ....
....

STATUTORY REQUIREMENTS

List any statutory requirements of exhaust systems relating to installation and functioning of the system when considering the following:

Condition, security and serviceability
....
....

Noise emission ....
....

Modifications ....
....

Chapter 6

Petrol (Carburettor) Fuel Systems

Petrol fuel system layouts (including fuel injection)   83
Mechanical fuel pump   84
The constant depression (variable choke) carburettor   84
Ford variable Venturi carburettor   85
Fixed-choke and constant depression carburettors   86
Fixed-choke carburettors   86
Air-bleed compensation system   86
Slow running   87
Sealed idle system   88
Cold starting   88
Strangler cold starting device   89
Acceleration devices   89
Plunger acceleration pump   89

Twin-choke carburettors   90
Progressive choke (two-stage) carburettors   91
Test equipment   92
Diagnosis: fuel system testing   93
Diagnostics   94
Fuel system protection during use   95
Routine maintenance   95
Mixtures, air–fuel ratios   96
Combustion   96
Fuel – technical terms   97
Fuel – statutory requirements   97
RON (research octane number)   98
Types of fuel used   98

PETROL FUEL SYSTEM LAYOUTS

The fuel supplied to spark-ignition petrol engines is by either carburettor(s) or fuel injection.

Identify the types shown below:

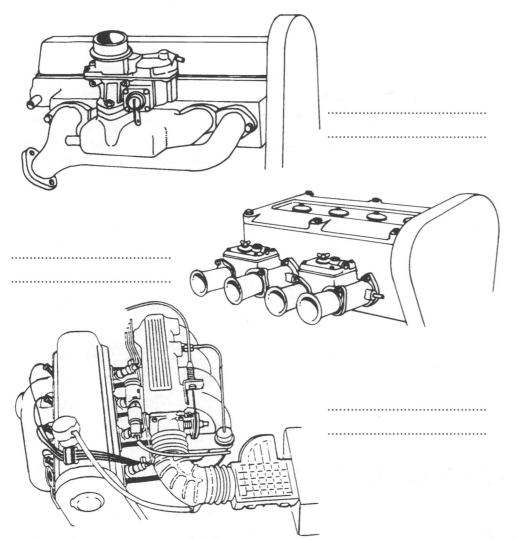

....

....

....

....

....

....

METHODS OF INTRODUCING FUEL INTO THE AIR STREAM

Show by arrows and describe the flow of air and fuel as it is drawn into the engine:

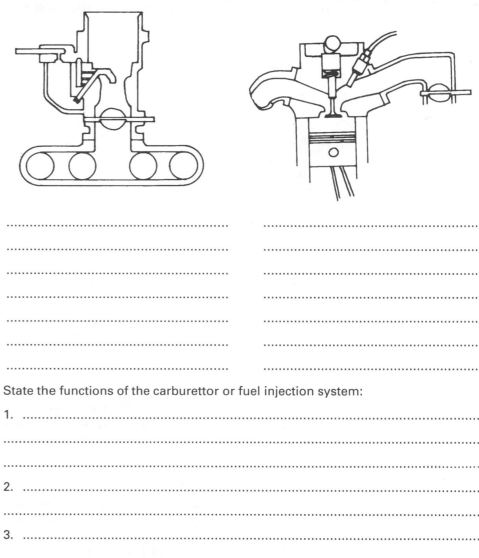

....

....

....

....

....

....

....

State the functions of the carburettor or fuel injection system:

1. ....

....

....

2. ....

....

3. ....

....

MECHANICAL FUEL PUMP

State the basic operation of the fuel pump shown.
Indicate the flow of fuel through the pump.

Outlet port   Inlet port

What happens to the pumping action when the carburettor is full?

....

....

....

List common faults that occur with mechanical pumps:

....

....

....

....

....

....

....

How would the more modern pump shown differ from the above type?

© ROVER

....

....

....

....

THE CONSTANT DEPRESSION (VARIABLE CHOKE) CARBURETTOR

One method of controlling the mixture strength is to vary both the choke area and the effective jet size in accordance with engine speed and load.

A simplified drawing showing the construction of a constant depression carburettor variable choke is shown below.

The piston assembly and throttle valve are in the positions occupied when the engine is idling, that is, the choke area and effective jet size are as small as possible to give the required idling speed.

Name the arrowed parts.

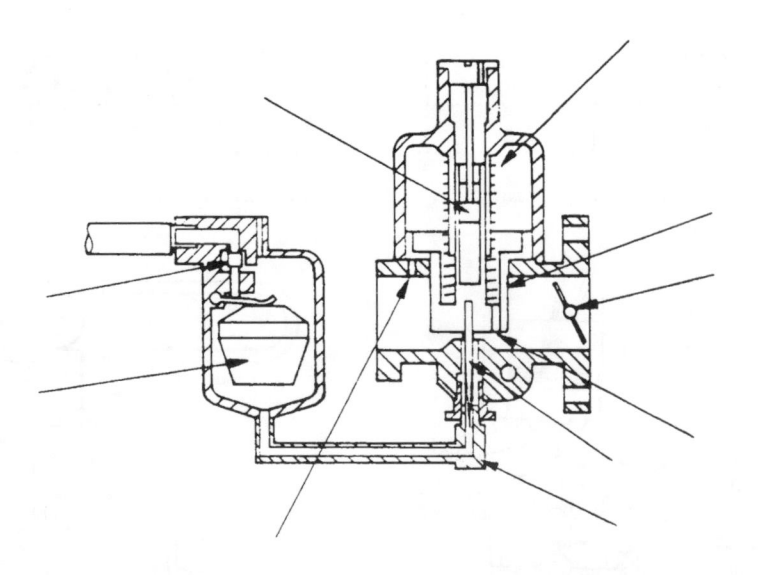

Why is this type known as a constant depression carburettor?

....

....

FORD VARIABLE VENTURI CARBURETTOR

The diagrams show a constant depression carburettor that incorporates a pivoting air valve operating a needle that slides horizontally in the jet.

Name the indicated parts on the views shown below:

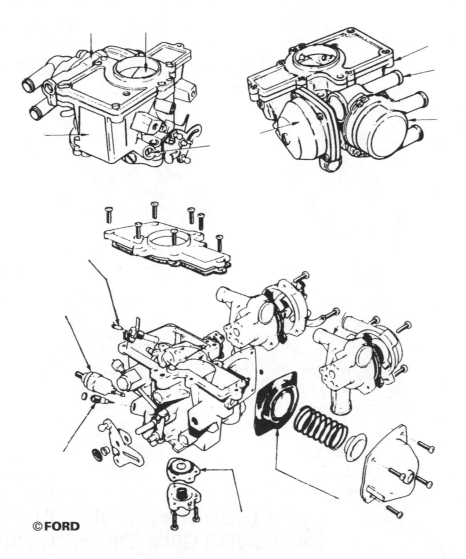

©FORD

Describe the basic auxiliary system of the carburettor:

....

....

....

....

....

....

....

....

....

The main air-flow system is shown below:

© FORD

Describe the control of air flow for the engine loadings stated:

Part-load position ....

....

....

Full-load position ....

....

....

85

FIXED-CHOKE AND CONSTANT DEPRESSION CARBURETTORS

Identify the carburettors below as either 'fixed-choke' or 'constant depression' types.

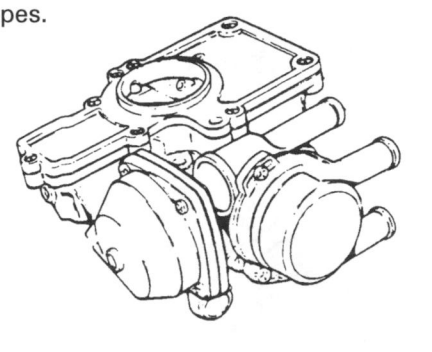

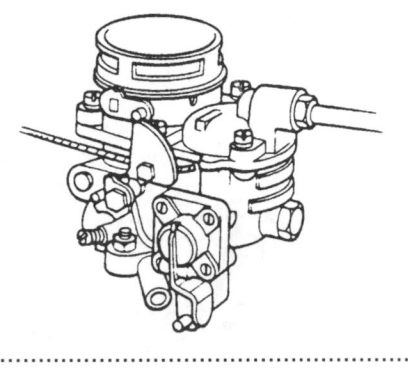

....

....

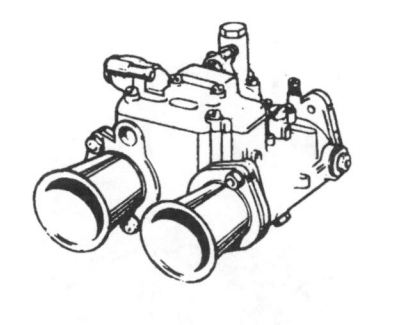

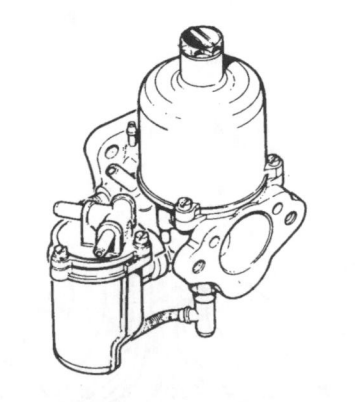

....

....

FIXED-CHOKE CARBURETTORS

The carburettor must control the quantity and proportion of fuel and air entering the cylinders to suit the engine speed and load requirement.

How is the fuel flow affected if the air flow is increased?

....

....

....

AIR-BLEED COMPENSATION SYSTEM

In this system, petrol is fed into a well which contains an 'emulsion tube' and an 'air correction jet'. With increase in engine speed, the fuel level in the well falls and progressively uncovers air-bleed holes in the emulsion tube. The extra air admitted as a result of this action prevents gradual enrichment of the mixture.

Examine a Solex type carburettor similar to that shown below and complete the drawing by naming the parts. Indicate also the petrol level in the system.

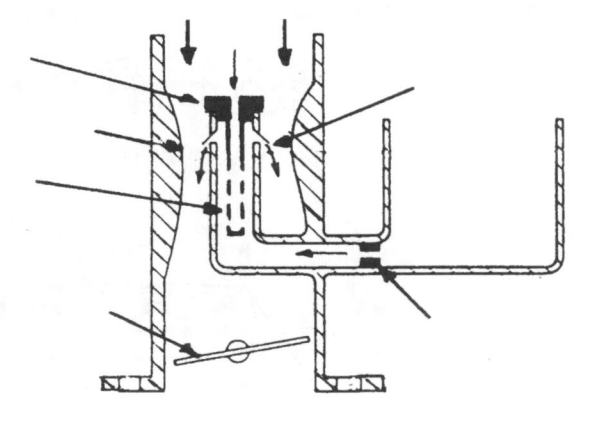

What is the function of the air correction jet?

....

....

What other function does the air-bleed perform?

....

....

....

How could the mixture strength and degree of correction be altered to suit different engines with this type of carburettor?

....

....

SLOW RUNNING

To allow the engine to run at low speeds, why is it necessary to use a mixture supply system which is completely separate from the main (and compensating) system?

....

....

....

....

The fuel is usually supplied via an orifice situated at the edge of the throttle butterfly as shown below.

Complete the sketch to show a Solex slow-running system. Add the idling air-bleed jet, pilot jet and volume control screw. Indicate the air–fuel mixture flow and name the arrowed parts.

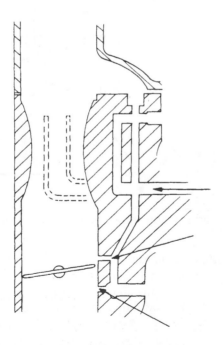

The operating principle of the slow-running system is that:

....

....

....

....

Why is a progression hole (or jet) necessary?

....

....

....

....

....

....

....

....

Investigation

Inspect a Solex carburettor similar to the one shown and if possible adjust the slow running on a similar carburettor.

Label on the drawing:

1. The throttle stop and volume control screw.

2. The cold start mechanism, including choke flap.

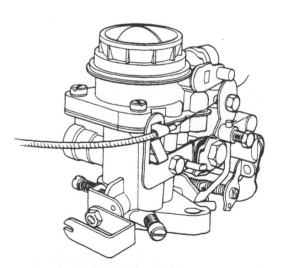

SEALED IDLE SYSTEM

Most modern carburettors incorporate an idle system in which the mixture control screw is sealed after setting and then should not require further adjustment. To increase or decrease engine speed, an air mixture screw should be turned. This screw is not found on pre-emission carburettors.

Name the parts indicated on the drawing and show the air and fuel flow through the system:

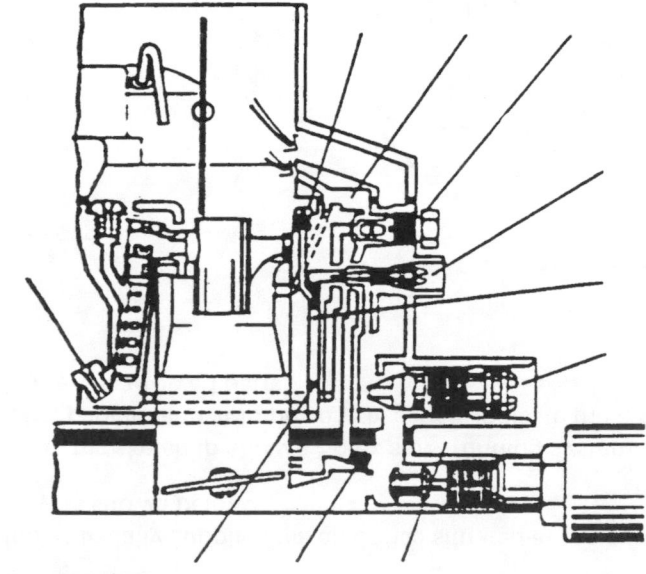

© VOLKSWAGEN

How is the fuel prevented from siphoning through the slow-running system when the engine is stopped?

....

....

....

The slow-running mixture by-passes the throttle valve, which must be fully closed during idling. The system can be described in the form of two separate layouts:

1. The basic mixture ....

....

....

....

2. By-pass mixture ....

....

....

....

....

....

....

....

COLD STARTING

When an engine is being started from cold, the carburettor must supply a richer mixture than is necessary under any other condition. To do this a separate starting device is employed; this may be either manual or automatic.

On the one shown (right) identify the type and indicate its position.

Type ....

Indicate the manual type on the drawing (left) and state how it is operated.

© VOLKSWAGEN

....

....

STRANGLER COLD STARTING DEVICE

The strangler valve is positioned in the air entry to the carburettor.

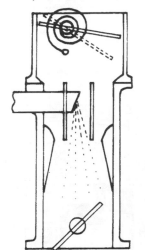

Identify the type of control unit shown at A and explain the function of the items numbered in B.

Type ....

A.

B.

© FORD

Function:

1. ....

2. ....

3. ....

Why is the spindle offset?

....

....

What is the function of the spring?

....

....

How is the strangler progressively opened as the engine warms up?

....

....

....

....

ACCELERATION DEVICES

To give the maximum power required for quick acceleration, the mixture reaching the cylinders should be slightly richer than normal. Without some form of acceleration device, rapidly opening the throttle would produce a weaker mixture than normal, that is, exactly opposite to what is required.

Why does a rapid throttle opening produce a weak mixture?

....

....

PLUNGER ACCELERATION PUMP

Name the parts of the pump assembly and describe its operation during both rapid and gradual operation.

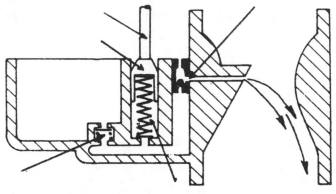

....

....

....

....

....

TWIN-CHOKE CARBURETTORS

There are many types of multi-barrel carburettors. The term 'twin-choke' is usually given to those carburettors whose throttles open simultaneously and have a single-float chamber supplying two separate but identical jet assemblies.

Below is shown a twin-choke Weber carburettor.

Examine the drawings to trace three of the main operating systems of the carburettor. Describe in each case the air–fuel flow.

Name types of engines that commonly use twin-choke carburettors:

....

....

State the advantages of this type of carburettor when compared with multi-carburettor installations:

....

....

....

....

....

....

....

© WEBER

Normal running:

....

....

....

Idle speed and progression:

....

....

....

Acceleration:

....

....

....

PROGRESSIVE CHOKE (TWO-STAGE) CARBURETTORS

Secondary barrel operated directly by a mechanical linkage.

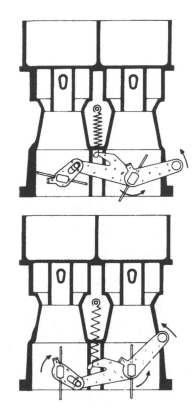

Identify the barrels and, for the throttle positions shown, complete the drawings by showing the petrol spray from the main petrol outlets and add arrows to the drawings to show the air flow through the barrels.

State the essential difference between a progressive and twin-choke carburettor:

....

....

....

In what circumstances are progressive choke carburettors considered necessary?

....

....

....

In the mechanical linkage type (left), when the primary linkage reaches three-quarter open position the secondary throttle commences to open and they both reach fully open position at the same time. During normal running, throttle action of the diaphragm type is similar. Describe the throttle action of the vacuum type when the engine is under load.

....

....

....

....

The sketch below shows a plan view of an inlet manifold for a progressive choke carburettor. Show the position of the carburettor barrels, when assembled, in relation to the induction tracts.

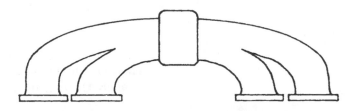

On the lower section of each sketch, show the throttle position related to the position of the linkage on the upper sketches.

Secondary barrel opened by vacuum operating a diaphragm mechanism.

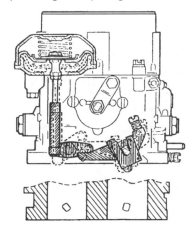

Throttle movement: normal acceleration air speed is sufficient to operate diaphragm.

© **SOLEX**

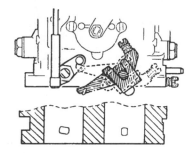

Acceleration under load: air speed to engine is low.

TEST EQUIPMENT

Examine the following types of equipment that are available for use and describe their operation, application and what care or maintenance is required during their use.

PRESSURE GAUGE

....
....
....
....
....
....
....
....
....
....

ENGINE TACHOMETER

Sketch multi-meter set at tachometer scale

....
....
....
....
....
....
....
....
....
....

EXHAUST GAS ANALYSER

....
....
....
....
....
....
....
....
....
....
....
....
....

Show sketch of workshop exhaust gas analyser.

VACUUM HAND PUMP

....
....
....
....
....
....
....
....
....
....
....

DIAGNOSIS: FUEL SYSTEM TESTING

Testing should be carried out in a logical sequence.
Tamper-proof seals are fitted over most modern carburettor adjustment screws. These must be removed before adjusting and renewed after adjustment is complete.

Describe preparatory checks that should be made before attempting any adjustments to the carburettor:

....

....

....

How should the engine be set up prior to checking?

....

....

....

....

Describe a typical idle speed and CO level test:
  e.g. Engine speed = 850 ± 50 rev/min, CO = 1.5 ± 0.25%.

....

....

....

....

....

Describe a typical fast idle speed check:
  e.g. Fast idle speed 1800 ± 100 rev/min.

....

....

....

....

Describe a method of carrying out the following fuel system checks:

Fuel consumption test

....

....

....

....

....

....

....

Fuel pump delivery pressure test

....

....

....

....

....

....

....

Carburettor float level check. Assume type where float is connected to lid.

....

....

....

....

....

....

DIAGNOSTICS: CARBURETTOR SYSTEM – SYMPTOMS, FAULTS AND CAUSES

State a likely cause for each symptom/system fault listed below. Each cause will suggest any corrective action required.

SYMPTOM/SYSTEM FAULTS	LIKELY CAUSE
Poor starting. Starter continually cranks engine without/before firing.	....
Engine seems low on power. It will not rev properly above a certain speed.	....
Engine runs roughly and stops when throttle is released.	....
Engine runs roughly and cuts out when revved.	....
When accelerator is depressed, the engine hesitates before picking up speed.	....
During over-run (foot off the throttle) popping noises occur in the exhaust. Sometimes the engine backfires.	....
Engine keeps running for a few seconds when ignition is switched off.	....
Car uses much more fuel than normal.	....
Garage smells of fuel. Fuel smells in car when driving.	....

FUEL SYSTEM PROTECTION DURING USE

Describe how the fuel system should be protected during use or repair from the following hazards:

(a) leakage ....

....

(b) contamination ....

....

(c) tampering ....

....

(d) overheating ....

....

(e) neglect ....

....

List the general rules/precautions to be observed while carrying out routine fuel system maintenance and running adjustments in respect of:

1. Avoiding fire hazards ....

....

....

2. Working in an enclosed space with engine running ....

....

....

3. Fuel in contact with skin ....

....

4. Disposing of waste materials ....

....

....

ROUTINE MAINTENANCE

State FOUR reasons for carrying out routine maintenance on the spark-ignition engine fuel system:

1. ....

....

2. ....

3. ....

4. ....

List typical Routine Maintenance/Adjustment checks:

1. ....

2. ....

3. ....

4. ....

5. ....

6. ....

7. ....

8. ....

9. ....

10. ....

Note: 'check' means also 'adjustment if required'.

Describe any special hand tools used to carry out routine maintenance and running adjustments:

....

....

....

....

MIXTURES, AIR–FUEL RATIOS

In order to obtain correct combustion, petrol and air must be mixed in correct proportions.
This ratio may vary depending on the type of running the engine is doing, but in view of economy and emission control must be kept within strict limits.

The chemically correct air–fuel ratio is ....
This is known as the Stoichiometric Rato.

10:1 would be .... 18:1 would be ....

Below is a graph showing air–fuel ratios for different car speeds.
The heavy line shows the results of a progressive increase in speed. The dotted lines show the fuel increase due to rapid acceleration at this speed.

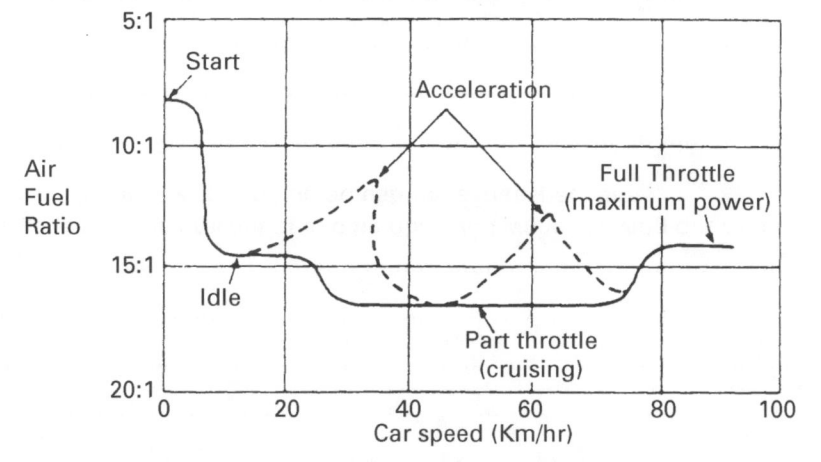

With the aid of the graph state typical air–fuel ratios for

Cold start	Economical running ....
Maximum power	Acceleration ....

The stoichiometric ratio of 14.7:1 by mass (14.7 kg of air to 1 kg of fuel) is given a lambda value of 1. This lambda value is also known as the Excess Air Factor.
Below is a table which shows possible engine running air–fuel ratios and their corresponding excess air factor values.

AIR–FUEL RATIO	11.76	13.23	14.7	16.17	17.64
EXCESS AIR FACTOR	0.8	0.9	1.0	1.1	1.2
	← Rich			Weak →	

COMBUSTION

The fuels used in heat engines are highly flammable liquids and when combined with the oxygen present in the air will burn when ignited.

How can the heat energy liberated during the burn be greatly increased?

....

....

....

The liberation of heat energy is known as ....

Complete the following statements with regard to the combustion of petrol.

The basic elements of all fuels are the elements ....
and .... When these are united with ....
(normally taken from the air) and ignited, heat is produced. Hence the term 'heat engines'.
The liquid hydrocarbon fuels used for motor vehicle engines are produced by distillation from crude oil. One such fuel is petrol which has a composition of about .... % of .... and .... % of ....

The fuel mixes with air which consists of 23% ....
and 77% ....

If the petrol and air mixture is burned in the chemically correct ratio of ....
the products formed by complete combustion will be:

1. ...., produced by combining the hydrogen in the fuel with the .... in the air;

2. .... , produced by combining the carbon in the fuel with the .... in the air;

3. .... , which is an inert gas.

When the engine is supplied with a slightly rich mixture, power output is usually .... This is because during combustion all the available .... in the air is used.

An undesirable result, however, is that some of the carbon will then burn to produce exhaust gas with an excess of ....
which is ....

FUEL – TECHNICAL TERMS

What is meant by the following terms with regard to fuels?

Ignition temperature ....

....

....

....

Lower and upper explosive limits ....

....

....

....

....

Volatility ....

....

....

....

Volatile fuels ....

....

Non-volatile fuels ....

....

Flash point ....

....

....

Spontaneous combustion ....

....

....

FUEL – STATUTORY REQUIREMENTS

Petrol presents a potential fire hazard and must, by law, be stored in a responsible and safe manner.

State the requirements for storing fuel in a workshop.

....

....

....

....

....

Fuel tanks fitted to vehicles must be corrosion resistant and must be able to withstand a pressure which is twice the operating pressure. Any over-pressure must automatically escape through a safety valve into the fuel vent system. Fuel must not escape past the filler cap.

Where should the fuel tank be positioned in a vehicle?

....

....

....

....

....

....

....

State the statutory requirements relating to the composition of exhaust gas emissions:

....

....

....

....

RON (RESEARCH OCTANE NUMBER)

The octane number of a fuel is a measure of the fuel's resistance to knock or detonate under pressure. RON refers to the method used to determine the octane number.

Explain how the octane rating of a fuel influences detonation.

....

....

....

....

LEADED FUEL

Before 1980 the most common additive in fuel to easily prevent detonation and so raise the octane number of fuel was tetra-ethyl-lead. Now leaded fuel is only available in the U.K. from service stations supplied by a company authorised to distribute leaded fuel nationwide. European legislation allows leaded petrol to be on sale for as long as there is a demand for it.
Why is leaded fuel not used in modern petrol engines?

....

....

LRP LEAD REPLACEMENT PETROL

This fuel is sold at all the major fuel service stations as a replacement for 4 star 98 RON leaded fuel. It contains potassium (10 to 12 mg\litre) as a lead substitute. If compression ratios are above 9 to 1 this may not be enough to protect the exhaust valve seats properly. An alternative to using this fuel in older vehicles is to use premium unleaded petrol and add a lead replacement substitute.

Name lead substitute products which use as a base the elements listed below.

LEAD REPLACEMENT SUBSTITUTE	PRODUCT
Phosphorus	
Manganese	
Potassium	
Sodium	

Note. Sodium can damage turbochargers. Never mix a manganese based substitute with a potassium based substitute and never overdose.

LEAD FREE FUEL

This is chemically a more complicated fuel. It is the recommended fuel for modern engines and engines able to be converted to lead-free fuel (that is, engines having valves and valve seats that are made from materials sufficiently hard, to prevent them from prematurely burning out in service).

List the grades of petrol that can be currently purchased.

....

....

SULPHUR IN FUEL

Sulphur is a natural element contained in all crude oil and while most is removed during the normal refining process further refining is required to obtain an ultra low sulphur content.

Sulphur in petrol was cut by regulation from 500 ppm to 150 ppm in the year 2000 and must be further reduced to 50 ppm by the year 2005.

To qualify for a fuel duty advantage the maximum fuel sulphur content is ....

What type of emissions does sulphur in fuel produce?

....

LPG LIQUEFIED PETROLEUM GAS

LPG is a by-product of the crude oil-to-petrol refining process.

What are the main constituents of LPG? ....

For safety what is added to LPG? ....

Many vehicles are being converted to use LPG, particularly taxis, police vehicles and local delivery vehicles in areas where LPG outlets are available.

LPG is normally used as a dual system. What is meant by this term?

....

....

The environmental benefits of LPG are that harmful exhaust products are substantially reduced when compared to petrol, e.g. carbon monoxide 75%, hydrocarbon 85% and oxides of nitrogen 40% less.
How does performance and economy compare to petrol?

....

....

....

....

Chapter 7

Diesel Fuel Systems

Diesel (CI) engine fuel system	100
Types of fuel injection pump	100
Compression-ignition fuel system layouts	101
Fuel tank	102
Low- and high-pressure fuel lines and unions	102
Lift pump, diaphragm-operated	102
Fuel filters	103
Primary filters	103
Cross-flow and down-flow filters	103
In-line fuel injection pump	104
Delivery valve	105
Governors – basic principles	105
In-line pump all speed mechanical governor	106
Minimec – leaf spring governor	106
Distributor-type pump	107
Bosch VE distributor pump	109
Injectors	110
Types of nozzle	111
Cold starting	112
Heater (glow) plugs	112
Diagnostics	112
Induction manifold heaters	113
Excess fuel device	113
Fuel line heaters	114
Fuel pumps – interrelationship with turbochargers	114
Fitting and timing injection pumps	115
Rotary pump timing	116

Fuel pump testing	117
In-line pumps – phasing and calibrating	117
Injector testing	118
Equipment for testing diesel fuel systems	119
Portable equipment	119
Portable testing equipment	120
Smoke emission test	120
Pressure-time injection system	121
PT fuel injector	122
Combined pump-injector	123
Electronic unit injector system	124
Electronically programmed injection control (EPIC) system	125
Electronic fuel pump control – EPIC 80	126
EPIC 80 pump – electronic operation	127
Component arrangement of the EPIC system	127
Common rail injection system	128
Ford – second generation – common rail system	128
Using an exhaust particulate filter	129
Pollution control	129
Fuel system protection during use	130
Diagnostics	131
Statutory requirments	132
Routine maintenance	132
Diesel fuel	133
Cetane number	133
Energy conversion	133

DIESEL (CI) ENGINE FUEL SYSTEM

There are many kinds of CI fuel supply systems, but at this introductory stage it is convenient to consider them as being of two basic types, those having in-line and those having rotary fuel injection pumps.

Identify the types of pumps shown below:

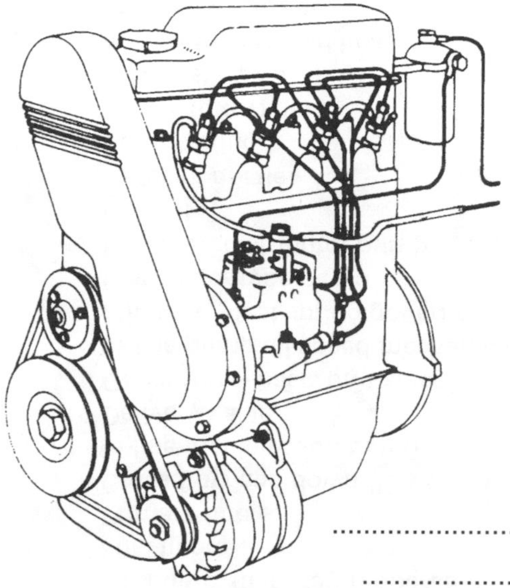

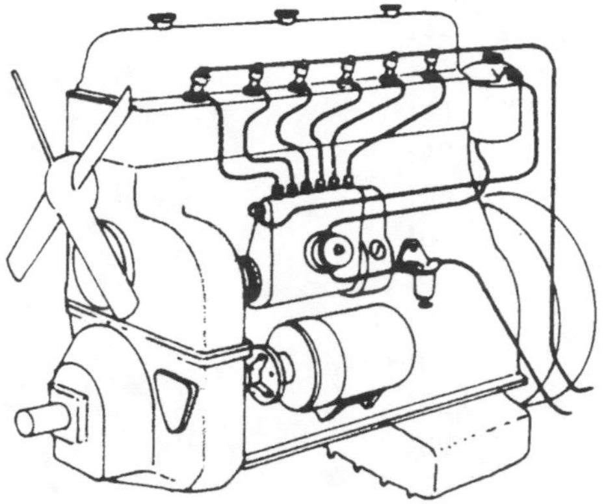

....

....

The purpose and function of a system is to provide:	Principal components involved:
Fuel supply to the engine	
Correct timing	
Correct quantities according to engine requirements	
Fuel in atomised form	
Clean fuel	

TYPES OF FUEL INJECTION PUMP

The rapid fuel pressure build-up to the point of injection is created by some form of jerk pump, usually an in-line or distributor type pump. In the first system shown below a constant pressure pump is used and the injectors are operated electronically.

Identify the common types of pump shown below:

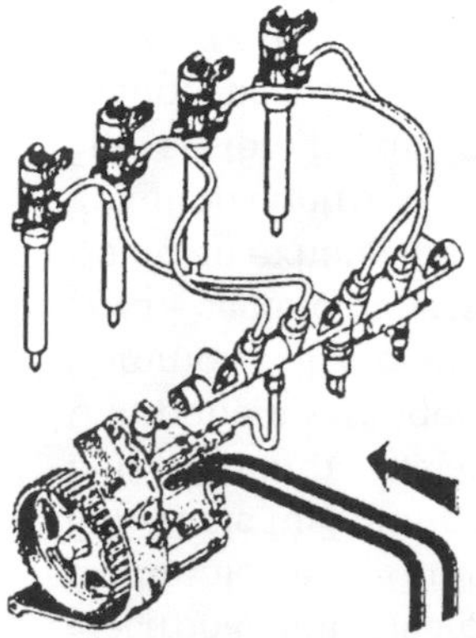

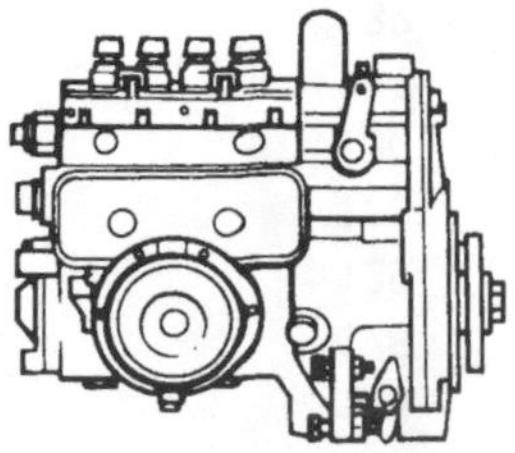

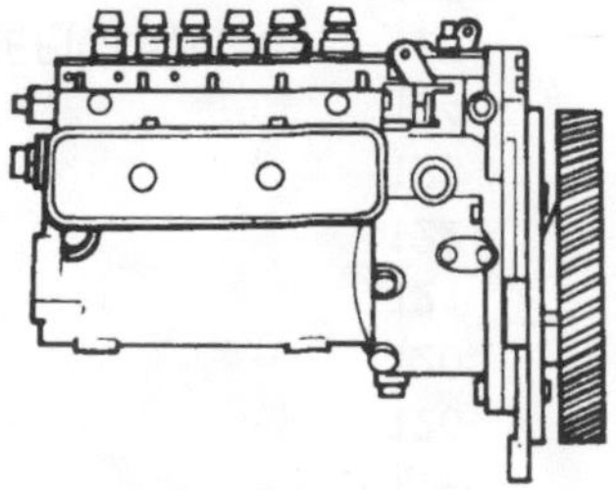

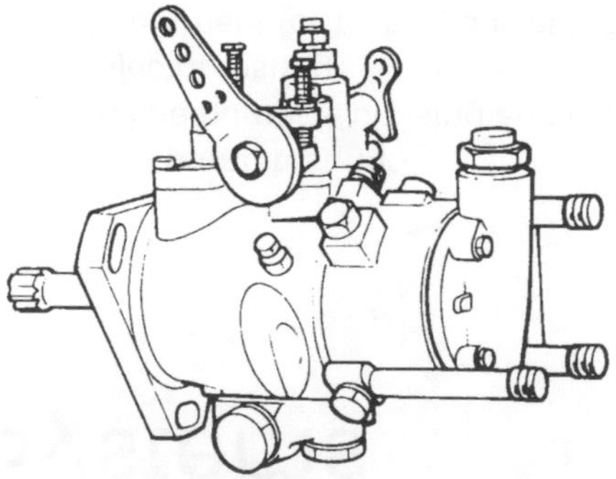

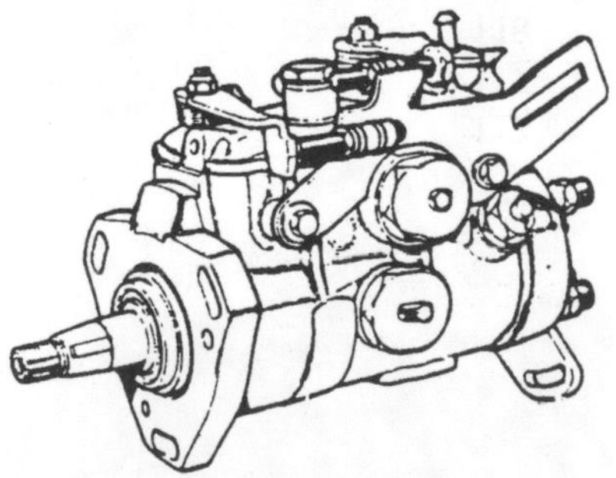

....

....

....

....

....

....

....

....

....

....

....

100

COMPRESSION-IGNITION FUEL SYSTEM LAYOUTS

The most common type of CI fuel system used on cars and on vehicles with engines of less than 3-litre capacity is a rotary or distributor type pump such as shown opposite. Larger CI engines may use a distributor type pump, but it is more likely that an in-line pump is used. These two designs of pump are the most popular types and are common on mass-produced vehicles. Some engine manufacturers produce their own fuel system design, examples being shown centre and lower right.

On all four drawings name the features indicated and show the direction of fuel flow.

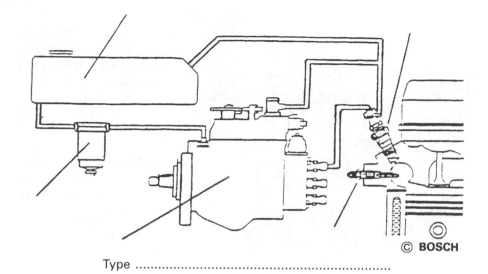

© BOSCH

Type ....

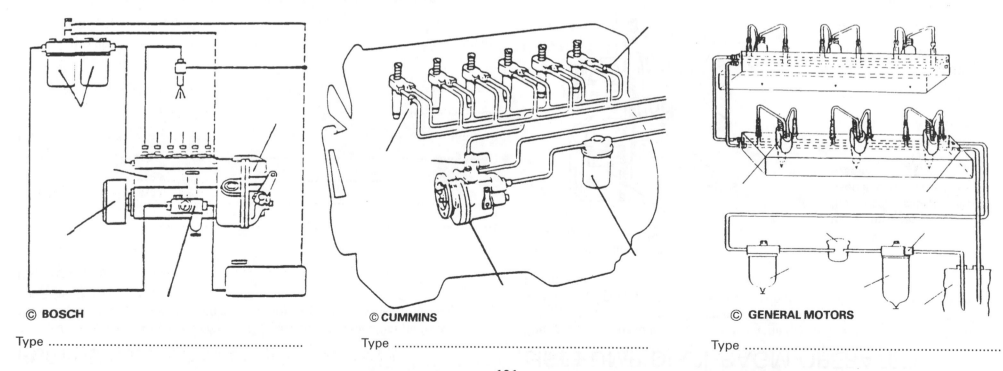

© BOSCH

Type ....

© CUMMINS

Type ....

© GENERAL MOTORS

Type ....

101

FUEL TANK

The fuel tank on a heavy commercial vehicle is considerably different from that fitted on a diesel-engined passenger car. Examine such fuel tanks and state their basic differences.

Heavy Commercial Vehicle ....

....

....

....

....

Diesel Engined Car ....

....

....

....

....

....

LOW- AND HIGH-PRESSURE FUEL LINES AND UNIONS

The low-pressure fuel lines on a CI engine may be of plastic similar to those used for petrol fuel pipes, or made from steel. The steel pipes look externally very similar to the high-pressure fuel lines.

Examine the union connections and sections or ends of high- and low-pressure fuel lines, and then with the aid of sketches show their constructional differences.

LIFT PUMP, DIAPHRAGM-OPERATED

The lift pump is used to supply fuel from the tank to the injection pump on the low-pressure side of the fuel system. It is basically similar to those used on petrol engines. The pump may be driven by the engine camshaft or by the injection pump camshaft.

Name the indicated parts of the lift pump shown below.

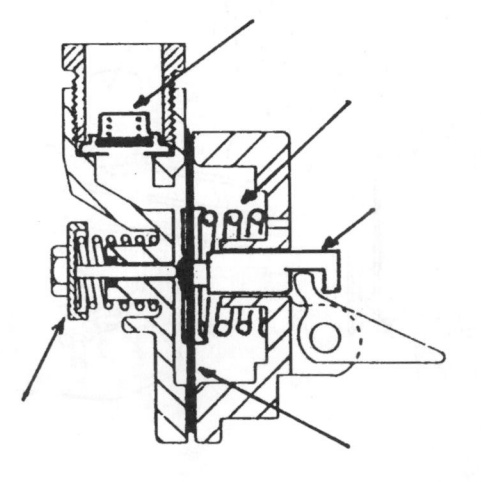

State TWO ways in which this pump may differ from the mechanical type of pump used in carburettor fuel systems.

....

....

....

....

....

....

....

HAND PRIMER

The diagram – right – shows a hand operated priming pump fitted on a Lucas/CAV system. The primer is mounted directly onto the fuel filter. Bosch and later type Lucas/CAV systems do not normally require priming. They are theoretically self bleeding.

© FORD

FUEL FILTERS

Why is it extremely important to filter the fuel minutely before it enters the injection pump?

....

....

PRIMARY FILTERS

These usually consist of a bowl which will allow large water droplets and particles of dirt to settle out. Describe the methods of fuel filtration shown.

Indicate the directon of fuel flow
in the sedimenter shown:

OUT   IN

An alternative or addition to the
sedimenter would be the fitting of a
filter agglomerator.

....

....

....

....

....

....

....

....

....

Show the direction of fuel flow.

AGGLOMERATOR

CROSS-FLOW AND DOWN-FLOW FILTERS

The main filter is normally of a paper element type (although fabric elements are still available).

Types of design determine whether the fuel flows down (or up) through the filter (commonly on CAV types) or across the filter similar to a conventional engine oil filter. How does the fuel filter differ from a conventional oil filter?

....

....

....

....

....

....

On the sketch below indicate the direction of fuel flow through the filter and name the main parts.

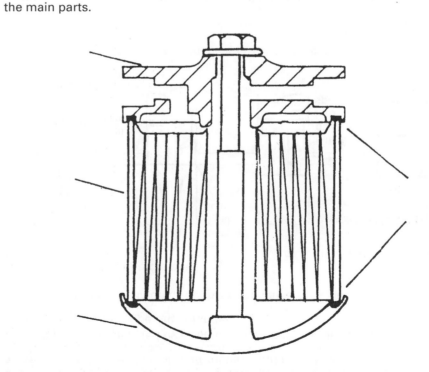

IN-LINE FUEL INJECTION PUMP

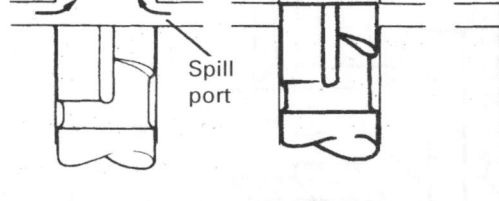

Most in-line multi-cylinder fuel injection pumps are castings which carry the pump camshaft in the lower portion and the pumping elements in the upper portion. Indicate these items.

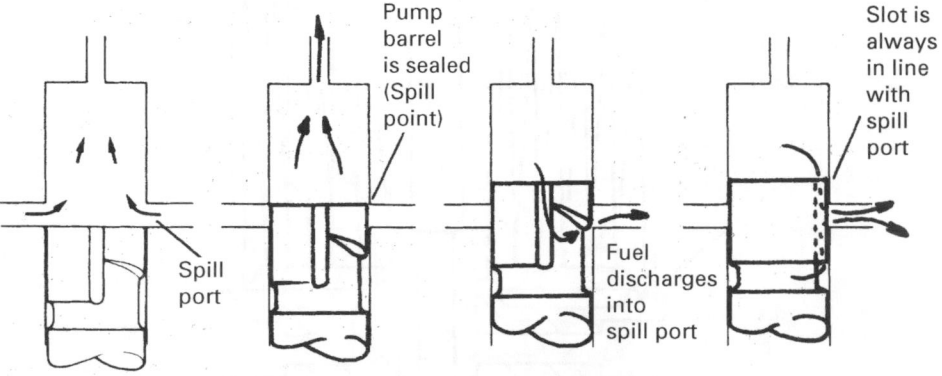

Below are shown two designs of pumping elements; both work on exactly the same principle, but some details are different.

Name the parts indicated and state how the quantity of fuel delivered is varied:

....

....

The injection pump must meter exact quantities of fuel, and deliver each charge at the correct degree of crankshaft rotation and at high pressure.

Each engine cylinder is fed by a separate pumping element consisting of a constant-stroke, cam-operated plunger.

Explain the plunger's action through a complete stroke, during acceleration and stopping.

Pump barrel is sealed (Spill point)

Slot is always in line with spill port

Spill port

Fuel discharges into spill port

Fuel filling   Delivery commences   Delivery ceases   Fuel cut-off

....

....

....

....

....

....

....

....

....

....

....

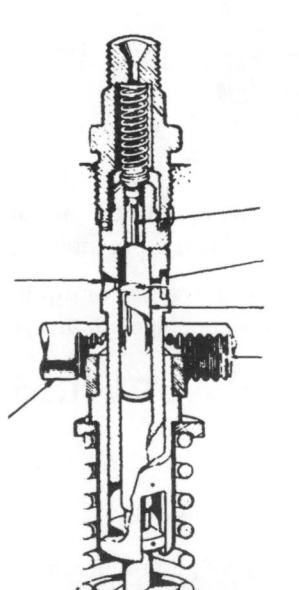

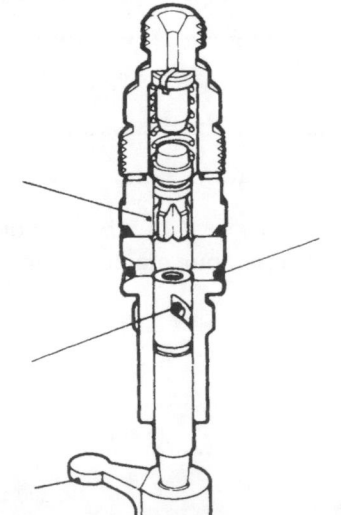

© LUCAS CAV

DELIVERY VALVE

This valve, or anti-dribble device, performs a dual function. It acts as a non-return valve and as a pressure unloading device. Explain this action.

....

....

....

....

Carefully examine a delivery valve, such as the one shown.
Indicate on the drawing the unloading volume.

The sketch shows a valve in its open position. Indicate the fuel flow.

State the effect of a worn delivery valve seat:

....

....

Explain the purpose of the 'unloading collar' machined on the delivery valve:

....

....

....

....

GOVERNORS – BASIC PRINCIPLES

A governor is fitted to all in-line and distributor pumps. In automobiles the engine speed is controlled by a combination of the governor's action and the driver's positioning of the accelerator pedal. When the accelerator is depressed the governor mechanism moves the control rod forward. This in turn rotates the plunger (or elements) to supply more fuel and increase engine speed. At a speed determined by the throttle position, the governor will operate and pull the rod back slightly to reduce the fuel supply and engine speed, as the speed reduces the control rod will be moved the other way to slightly increase the speed.

State the basic functions of the governor:

1.   ....

2.   ....

3.   ....

The governor must also provide a setting for excess fuel when starting, and be sensitive to torque controls and turbocharging pressure when these devices are fitted.

Complete the block diagrams below to show the governor functions.

| Diesel Engine |

| In-line pump |

| Govenor |

IN-LINE PUMP ALL SPEED MECHANICAL GOVERNOR

All mechanical governors rely on the action of centrifugal force attempting to throw a set of spinning weights outwards from their normal position and in doing so, overcoming some form of spring resistance.

MINIMEC – LEAF SPRING GOVERNOR

The type of governor described on this page is commonly found on high-speed diesel engines which may have a cubic capacity of up to 1.5 litres per cylinder.

The governor governs all speeds between idling and maximum speed. Stop screws control the position of the governor at idle and maximum speed.

Label the main parts in the sectional view of the pump below:

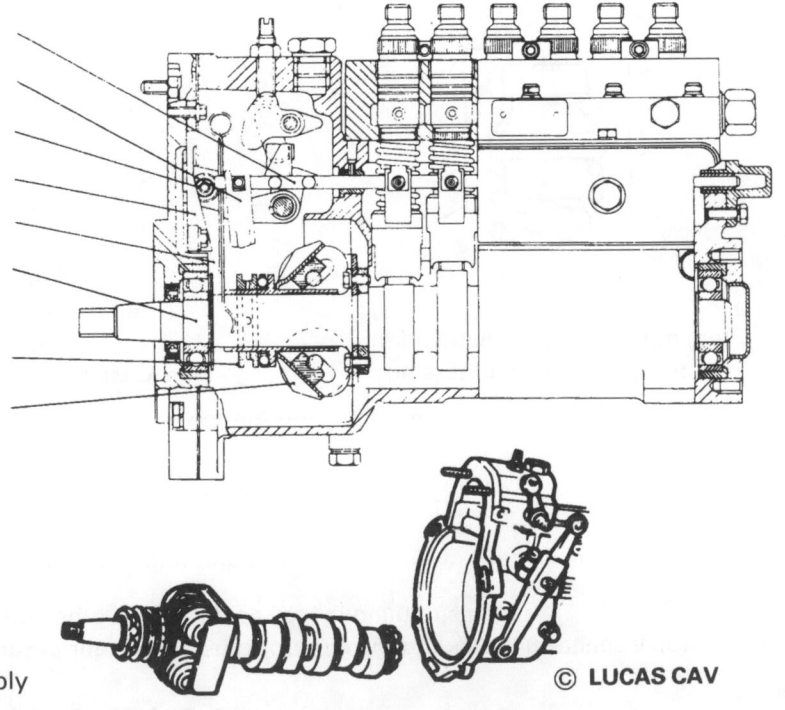

Camshaft and weight assembly

© **LUCAS CAV**

Examine an in-line pump.

Note (1) the operation of the roller weights and (2) the action of the leaf (governor) spring when the throttle lever is operated.

Explain the operation of the governor and name the important parts on the drawings:

....
....
....
....
....
....
....
....
....
....
....
....
....
....
....
....
....
....
....
....
....
....

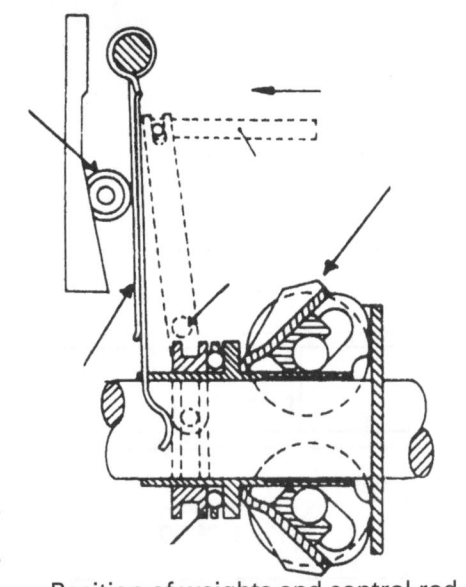

Position of weights and control rod when stopped and at maximum speed

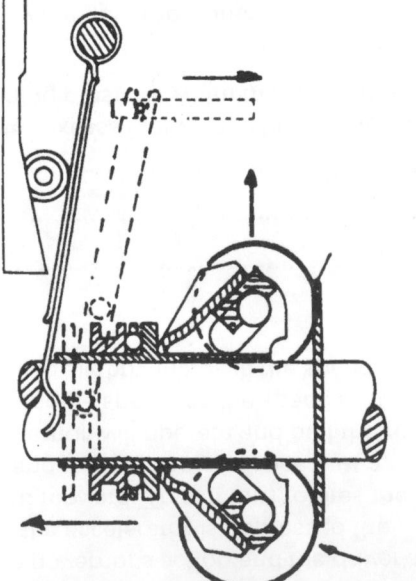

Position of weights and control rod when pump is operating at idle speed

DISTRIBUTOR-TYPE PUMP

The distributor-type pump has one common pressurising and metering device.

The pump plungers are operated by a precisely machined cam ring and the rotor unit is driven by the engine.

Complete the diagram below by adding arrows indicating the direction of fuel flow. Also on the lower diagrams sketch the position of the plungers.

With the aid of arrows show the flow of fuel through the pump and name the parts indicated.

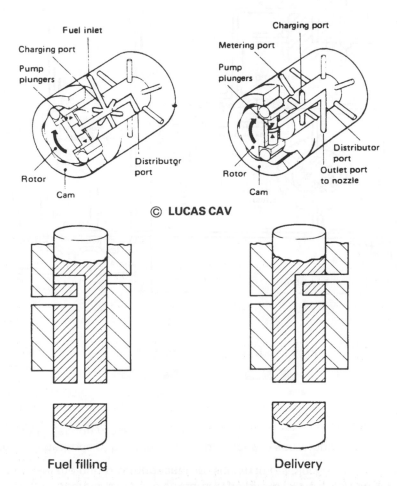

© **LUCAS CAV**

Fuel filling

Delivery

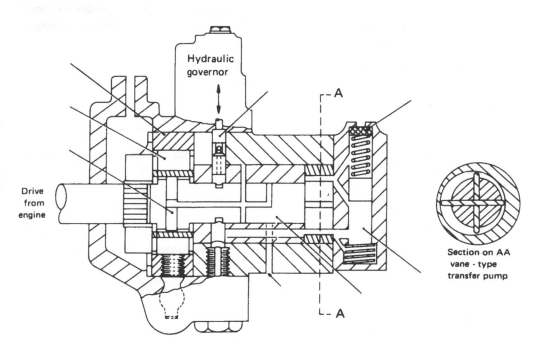

Section on AA
vane - type
transfer pump

Explain the purposes of the regulating valve:

....

....

....

....

....

....

....

The pump shown on the left is designed to operate an engine having ....
cylinders.

The principle of operation of the distributor-type pump is shown on the schematic diagram below; the pump components are shown in the block.

Use different methods of hatching on the pipelines to show input, transfer, metering and injection pressures.

input　　　　transfer　　　　metering　　　　injection

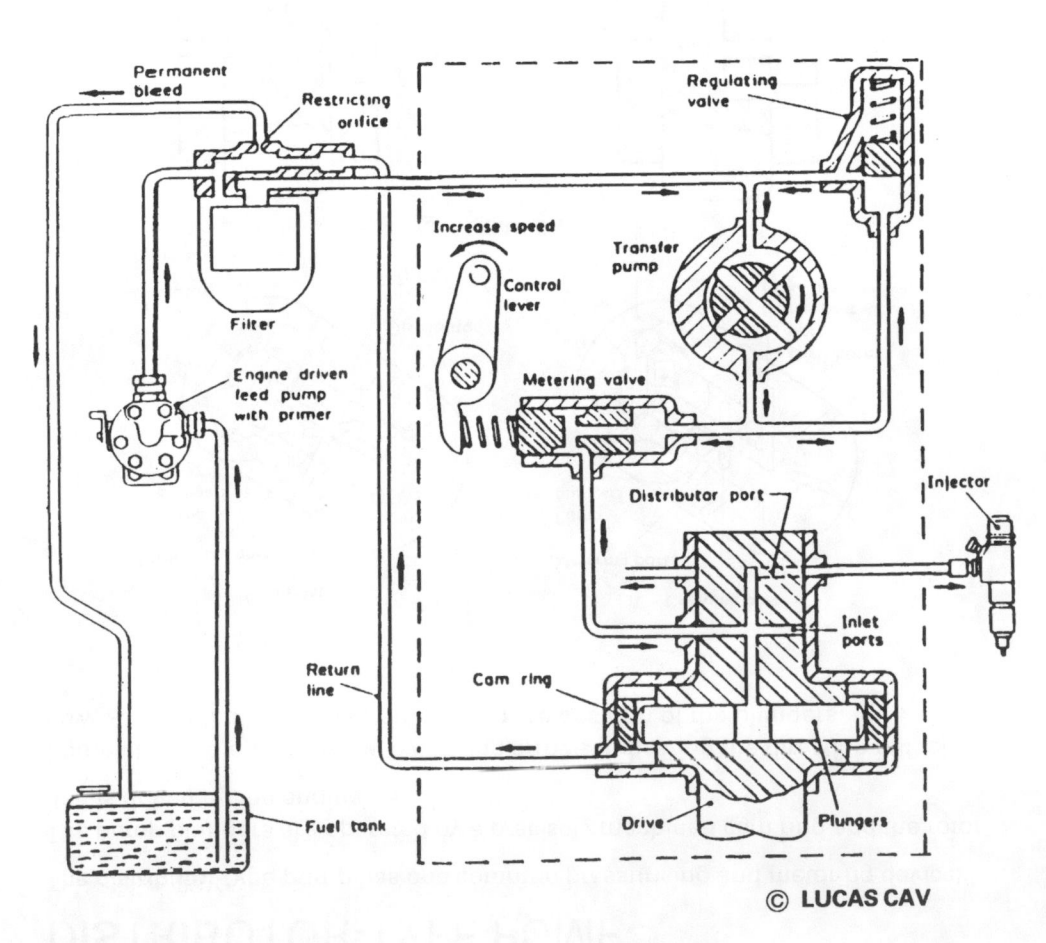

© **LUCAS CAV**

With the aid of the diagram describe the operation of the pump:

....

....

....

....

....

....

....

....

How are the moving parts lubricated?

....

....

....

State how this type of pump ceases to deliver fuel when it is desired to stop the engine:

....

....

State what maintenance this type of pump requires:

....

....

....

....

Note: The diagrams of the rotary pumps shown indicate the basic principles of such pumps. They do not show the extra complicating features incorporated into modern emission control type pumps.

BOSCH VE DISTRIBUTOR PUMP

This distributor pump is similar in many ways to the DPA pump described on the previous two pages.

Inspect the sectioned drawings of the pump below, name the features indicated and describe their function:

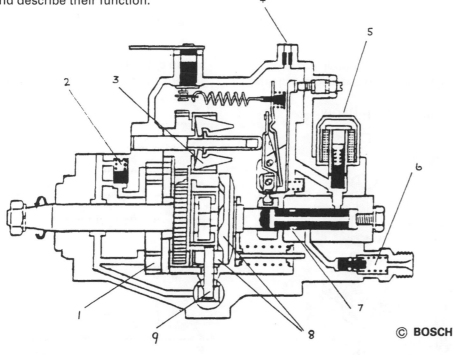

© BOSCH

1. ....
....

2. ....
....

3. ....
....

4. ....
....

5. ....
....
....
....

6. ....
....
....

7. ....

8. ....

9. ....

The distributor plunger moves in a rotary-reciprocating motion as it opens and closes the ports in the distributor head.

With the aid of the diagrams, describe the plunger fuel delivery action:

A. ....
....
....

B. ....
....
....
....
....
....

C. ....
....
....
....
....

A plunger   fuel entry

metering slit

high pressure chamber

B

governor control spool

to injector

C

fuel returns to governor housing cavity

© BOSCH

INJECTORS

The function of the injector is to deliver fuel in atomised form into the combustion chamber of the engine. The injector normally consists of a body and nozzle assembly.

Describe the fuel flow through the injectors:

....
....
....
....
....
....
....
....
....
....
....

State any special precautions that should be taken when fitting a nozzle nut to the injector body:

....
....
....
....

Name the parts indicated on the injectors shown.
Note each injector's method of fixing into the cylinder head.

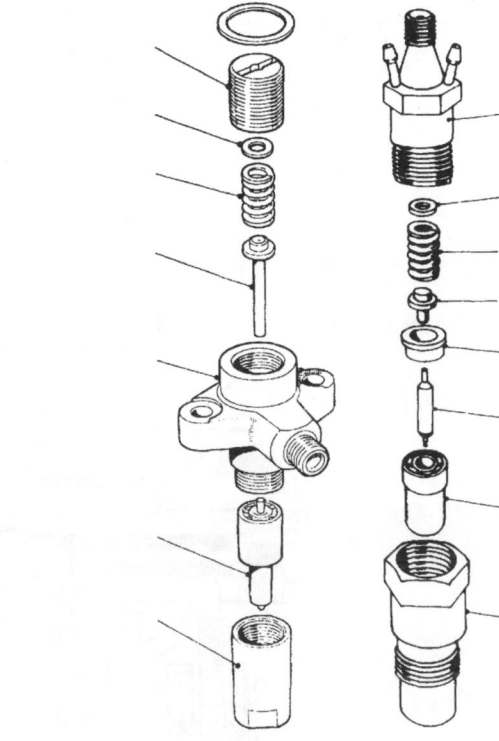

MULTI-HOLE   PINTLE

MULTI-(4)HOLE

STANADYNE
'SLIM-TIP'
(5)HOLE

Explain the purpose of the leak-off facilities provided on an injector:

....
....
....
....

Explain the meaning of needle lift:

....
....
....
....

Why does the injector buzz when operating?

....
....
....
....

TYPES OF NOZZLE

Different designs of combustion chamber demand special nozzles. The two main designs in common use are the hole type and pintle type.

The type of engine normally using a:

1. hole-type injector is ....

2. pintle-type injector is ....

Hole Type

These have an internal needle exposing from one to several holes.

Why is the long-stem nozzle the most popular type?

....

....

....

....

The sketches below show hole-type nozzles. Name the arrowed parts.

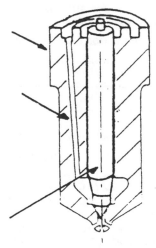

Multi-hole

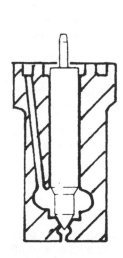

Single-hole

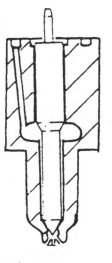

Long-stem

Pintle Type

This type gives a hollow cone-shaped spray. It is possible slightly to delay the delivery of the main spray by a modification to the pintle, thus giving a more gradual pressure rise and smoother running. Why is the pintle design used for indirect engine combustion chambers?

....

....

....

The sketches below show pintle-type nozzles.

PINTLE

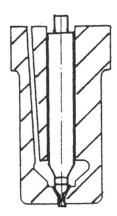

DELAY

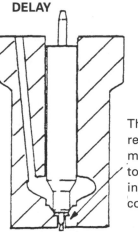

The pin requires more time to lift before injection commences.

PINTAUX

Show the fuel flow through the pintaux nozzle when the engine is cranking:

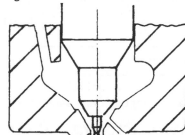

Explain the purpose of the pintaux-type of nozzle:

....

....

....

....

....

....

COLD STARTING

Cold starting compression-ignition engines can be a problem because of initial heat losses. This is particularly so with the indirect injection type. The difficulty can be overcome in several ways.

HEATER (GLOW) PLUGS

At the cylinder end these plugs have either a small tube or a coil of wire which glows at red heat when switched on. How long are the plugs energised?

....

....

....

The glow plug control unit governs the preheating period. The unit monitors the engine temperature via the coolant temperature sensor and will alter the pre-heat time to suit the conditions. Turning the ignition key to the second position most commonly triggers the preheating cycle. On some vehicles opening and closing the driver's door triggers the system.

Examine a heater plug circuit and complete the wiring diagram in the space below.

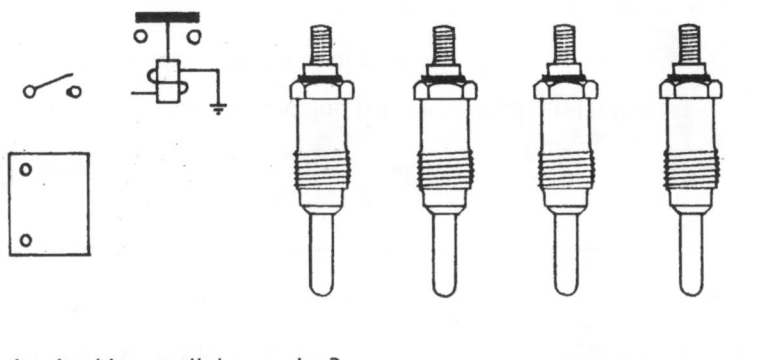

Is the circuit wired in parallel or series? ....

State the voltage power rating of each heater plug ....

State the amperage of the fuse in the circuit ....

DIAGNOSTICS: DIESEL ENGINE GLOW PLUGS – FAULTS AND CAUSES

State likely causes for each fault listed below. Each cause will suggest any corrective action required.

FAULT	LIKELY CAUSE	SHEATH
Melted or broken sheath		
Corroded sheath		
Open circuit where the sheath appears to be undamaged		

Note. When fitting glow plugs, to prevent over tightening, a torque wrench should always be used.

INDUCTION MANIFOLD HEATERS

These are fitted in the inlet manifold.

Describe a type that consists simply of a coil of resistance wire.

....

....

The most common form of manifold heating is to use a hot wire-coil which directly operates a fuel supply. Describe the action of the manifold heater shown below.

....

....

....

....

....

Name the parts of this manifold heater incorporating a fuel vaporiser.

Explain how the fuel is stored and supplied for use in the manifold heater unit.

....

....

....

EXCESS FUEL DEVICE

Most in-line pumps are fitted with a device which when actuated allows the control rod to move to a position in which excess fuel for easy starting will be delivered.

The push-button type is the simplest arrangement.

The sketch below shows the plunger in its non-operational position.

Name the parts indicated and describe the unit's operation as the engine is started.

....

....

....

....

....

....

....

....

....

....

Alternative types which are 'cheat proof', that is, the button cannot be 'tied down', may be used when required.

What are the statutory requirements regarding the use and location of excess fuel and cold starting controls?

....

....

....

....

....

....

FUEL LINE HEATERS

At very cold temperatures, wax crystals start to form in the fuel. This can be contained by placing additives in the fuel and this is done automatically in winter by the fuel suppliers. When it is known that the temperature is to be permanently cold (below – 9°C) fuel line heaters should be fitted.

Below is shown a typical fuel line heater.
The heating element is in the flexible pipe.

The fuel filter shown is connected to the engine cooling system. Indicate the cooling system connections.

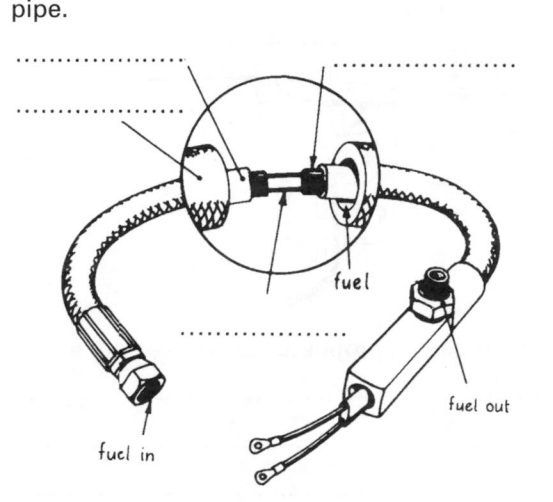

fuel

fuel out

fuel in

In the UK, since temperatures rarely drop below – 9°C, it is considered that by achieving good design and using additives in winter there is little need for the fuel to be heated.

List good design features which protect the fuel from excessive cooling:

....

....

....

....

....

....

FUEL PUMPS – INTERRELATIONSHIP WITH TURBOCHARGERS

If engines are turbocharged, what modification to the fuel pump is required?

....

....

....

....

....

....

....

....

....

© **LUCAS CAV**

Boost controller shown on sectioned DPC pump

The Lucas CAV DPC pump is a more advanced DPA-type (Roto diesel) design and incorporates features such as a solenoid fuel cut-off and high-pressure delivery valve similar to those shown on page 109.

State ways in which fuel injection pumps may be driven from the engine:

In-line pumps ....

....

....

Rotary pumps ....

....

....

....

114

FITTING AND TIMING INJECTION PUMPS

Describe a typical removal procedure for when it is necessary to fit an exchange pump.

Engine make .... Type of pump ....

....

....

....

....

....

....

....

....

....

....

....

....

....

....

....

....

....

....

....

....

When might it be considered necessary to fit a replacement fuel pump?

....

....

....

Both the in-line pump and the distributor-type pump must be fitted to deliver fuel at an exact point of crankshaft rotation. Most distributor-type pumps have a master spline to ensure correct timing. The in-line pump normally has a timing mark which must be in the correct position when coupled to the engine. The flywheel is usually very accurately marked with timing information.

Examine the specifications of several engines using in-line pumps and record below the static timing for commencement of injection.

Make and model			
Static timing in line pumps	°btdc	°btdc	°btdc

Describe a standard engine/pump timing procedure for an in-line pump:

....

....

....

....

....

....

....

....

....

....

....

....

....

ROTARY PUMP TIMING

Describe a rotary pump static timing procedure for a BOSCH fuel injection pump and an engine that requires top dead centre (tdc) to be found by removing a blanking plug from above a cylinder, inserting a probe and setting up a dial test indicator (DTI) to measure the probe movement.

ENGINE MAKE .... **PUMP TYPE** ....

POSITIONING ENGINE TO tdc

....

....

....

....

....

....

OBTAINING TIMING ON PUMP

....

....

....

....

....

....

....

....

© BOSCH

POSITION OF DTI ON BOSCH PUMP

....

....

....

....

....

....

....

....

....

Describe a pump static timing procedure for a CAV Roto-diesel fuel injection pump and an engine that requires top dead centre (tdc) to be found by inserting a pin for location.

The Roto-diesel pump is identified by two large bolts on the side of the pump.

ENGINE MAKE .... **PUMP TYPE** ....

POSITIONING ENGINE TO tdc

....

....

....

....

....

....

OBTAINING TIMING ON PUMP

....

....

....

....

....

....

....

....

© LUCAS CAV

POSITION OF DTI ON CAV PUMP

....

....

....

....

....

....

....

....

NOTE. When you carry out a similar workshop procedure on an engine, the manufacturer's specific instructions must be used.

FUEL PUMP TESTING

It is extremely critical for the efficient performance of the compression-ignition engine that the injection pump maintains precise deliveries of exactly metered quantities of fuel.

Before fitting to the vehicle the pump must be tested at various speeds to check and adjust the fuel output (calibration) and point of injection (phasing), and the maximum and minimum fuel supply (controlled by stops).

Below are shown two typical pump test stands. Describe their main features.

HARTRIDGE 600

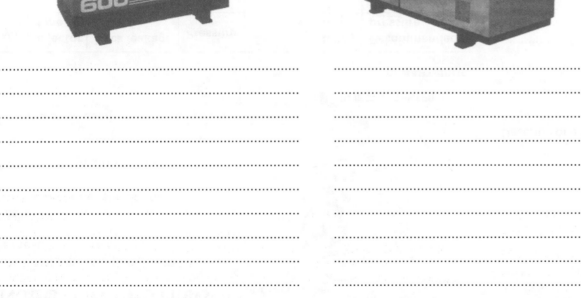

HARTRIDGE 2500

....
....
....
....
....
....
....
....

....
....
....
....
....
....
....
....

IN-LINE PUMPS

PHASING

What is the function of this adjustment?

....
....
....
....

CALIBRATING

What is the function of this adjustment?

....
....
....
....

State the effect of running a compression-ignition engine with the pump elements out-of-phase.

....
....

Explain the meaning of the term 'spill cut off'.

....
....
....

State the effect of a pump delivering unequal quantities of fuel to each of the engine cylinders.

....
....
....

INJECTOR TESTING

This should include pressure tests, leak tests and spray form.

Pressure-test different types of injectors and record observations below.

Make of test equipment ....

Model ....

Type of test oil used ....

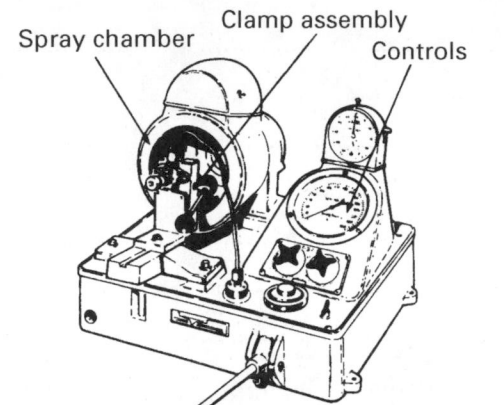

Spray chamber — Clamp assembly — Controls

INJECTOR TESTER

© **HARTRIDGE**

No.	Injector type	Back leakage time(s)	Pressure	Seat tightness	Spray form	Recommended pressure
1						
2						
3						
4						

State observations and recommendations in respect of each injector tested:

1. ....

2. ....

3. ....

4. ....

When might it be considered necessary to fit a replacement set of injectors?

....

....

Describe one method of adjusting the pressure setting of an injector:

....

....

....

....

....

....

Describe one method of quickly locating a faulty injector while *in situ* on the engine:

....

....

....

....

....

....

....

Using a pressure tester, observe the spray form of a hole-type and pintle nozzle. Sketch the correct spray outlines below.

EQUIPMENT FOR TESTING DIESEL FUEL SYSTEMS

Layout shows a very comprehensive diesel fuel system reconditioning test shop and also the type of equipment that is available from 'Hartridge Diesel Test Equipment'.

Why is diesel test equipment kept in a room of this type?

....

....

Name the equipment indicated on the drawing.

PORTABLE EQUIPMENT

When diagnosing engine/fuel system faults on a vehicle, small easy-to-use portable equipment is required.

TACHOMETER

Below are shown digital optical tachometers. Describe how they operate.

© HARTRIDGE

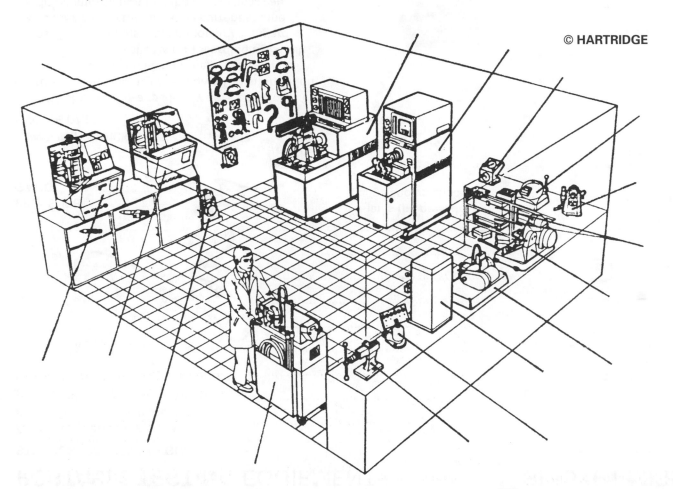

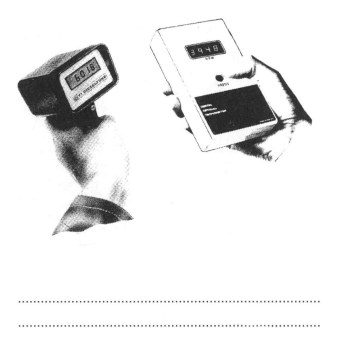

....

....

....

....

....

....

PORTABLE TESTING EQUIPMENT

STROBOSCOPIC TIMING GUN

A digital timing gun is shown below. To where is it fitted and what information can it provide?

....
....
....
....
....
....
....
....
....

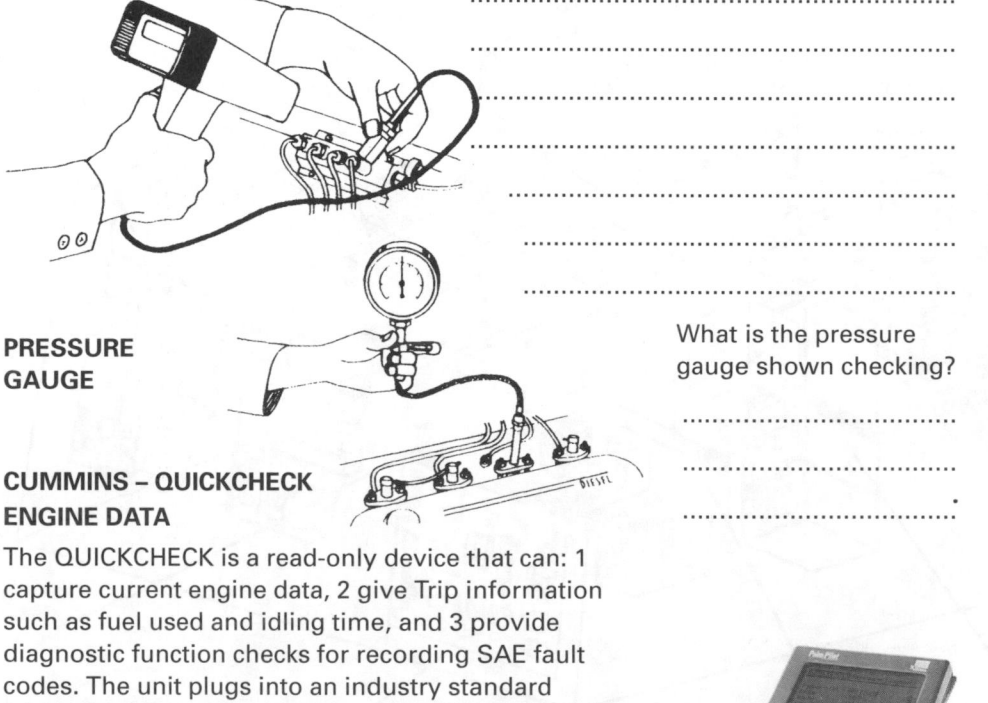

PRESSURE GAUGE

What is the pressure gauge shown checking?

....
....
....

CUMMINS – QUICKCHECK ENGINE DATA

The QUICKCHECK is a read-only device that can: 1 capture current engine data, 2 give Trip information such as fuel used and idling time, and 3 provide diagnostic function checks for recording SAE fault codes. The unit plugs into an industry standard SAE J1587 connector. State the current engine condition information that the quick check can capture.

....
....
....
....
....

© CUMMINS

SMOKE EMISSION TEST

A smoke meter must be used if the density of smoke is to be measured accurately.

Diagram shows a Hartridge smoke meter mounted on a portable cabinet connected to a vehicle being tested on a chassis dynamometer; there it can measure accurately the amount of smoke produced under different conditions of load, speed and power output.

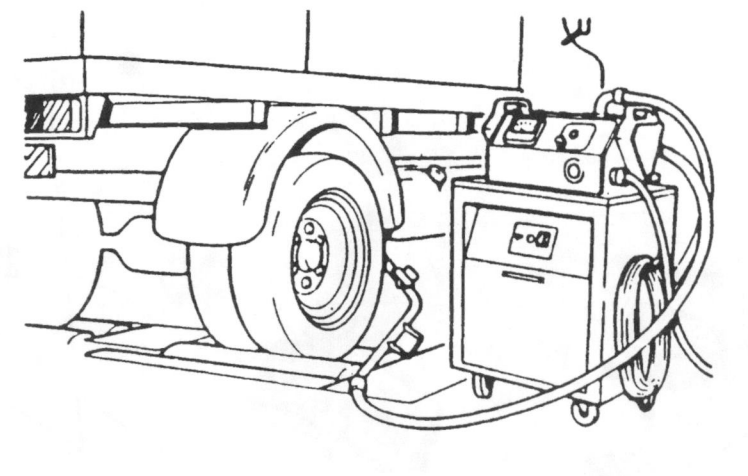

Describe how such a meter is used:

....
....
....
....
....
....
....
....
....

PRESSURE–TIME INJECTION SYSTEM

This system was developed by the Cummins Engine Company. Its operating principle is based on the fact that changing the *pressure* of liquid flowing through a pipe changes the amount coming out of the open end (that is, injector); the length of *time* during which the injector is 'open' is controlled mechanically. Hence the name, pressure–time system.

The basic mechanical linkage to operate the injector is shown below.

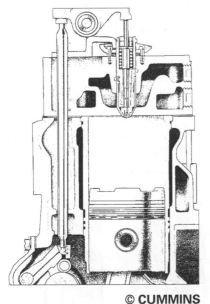

© CUMMINS

With this system about 80% of the fuel delivered to the injectors is returned to the tank. What are the reasons for this?

....

....

....

....

....

How are the high pressures needed to inject the fuel into the combustion space created?

....

....

....

....

....

....

....

....

....

....

....

The drawing below shows the basic layout of the Cummins PT fuel line system. Name the parts indicated, and use arrows to indicate the direction of fuel flow.

© CUMMINS

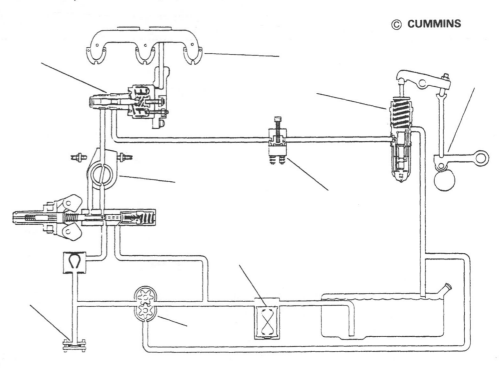

Describe how the fuel is supplied to the injector:

....

....

....

....

....

....

PT FUEL INJECTOR

The drawings below show the nose of a PT injector at various stages of operation.

Explain what is occurring at each stage.

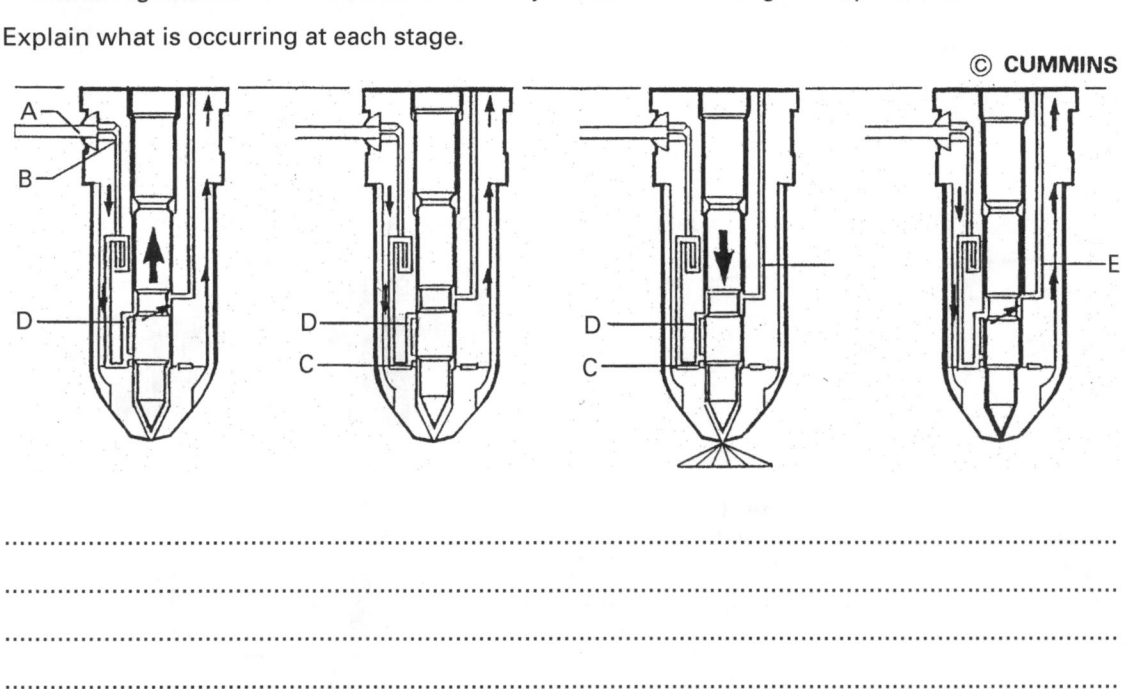

© CUMMINS

....
....
....
....
....
....
....
....
....
....
....
....
....

Diagram shows the injector layout on the Vee-8 engine. Indicate the direction of fuel flow and name the main parts.

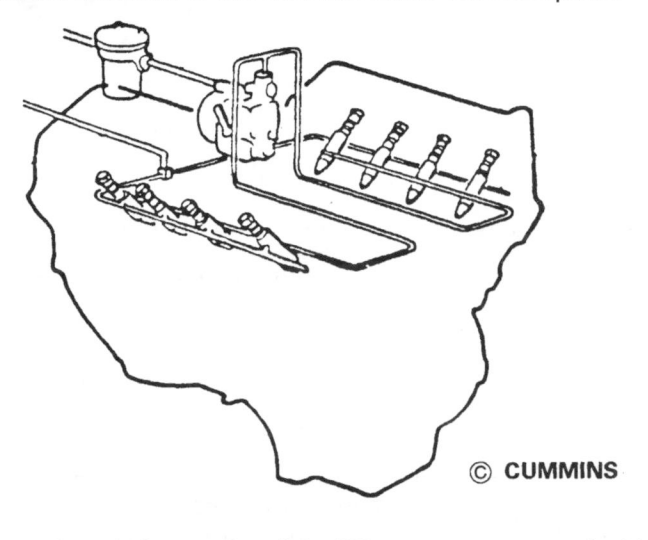

© CUMMINS

What are the relative merits of the PT systems compared with the more conventional fuel injection systems?

ADVANTAGES

....
....
....
....
....
....

DISADVANTAGES

....
....
....
....

COMBINED PUMP–INJECTOR

With this system the injector takes on the dual function of metering and injection. It is commonly known as the General Motors system.

For layout of system see page 101.

Name all the important parts of the combined pump–injector unit shown. Indicate the direction of rack movement to increase and decrease fuel.

Describe the basic operation of the system:

....

....

....

....

With the aid of the diagrams explain the injector operation. © **GENERAL MOTORS**

The closed type shown below is an alternative injector nozzle.

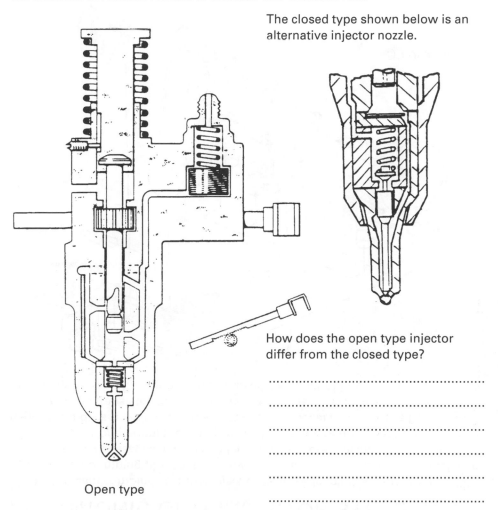

Open type

How does the open type injector differ from the closed type?

....

....

....

....

....

1.　Top of stroke　Lower port

2.　Start of injection stroke

3.　End of injection stroke

4.　Bottom of stroke　Upper port

....

....

....

....

....

....

123

ELECTRONIC UNIT INJECTOR SYSTEM

The EUI system combines the fuel injection pump and injector in a single element. The pumping plungers are driven by the engine camshaft, and fuel feed and spill are by passages in the cylinder head. Sensors provide information to the ECU on engine operation; they indicate coolant temperature, boost pressure, engine speed, camshaft position and accelerator pedal position. The ECU compares and translates the signals and then actuates each injector solenoid to open at the correct time and deliver a specific amount of fuel.

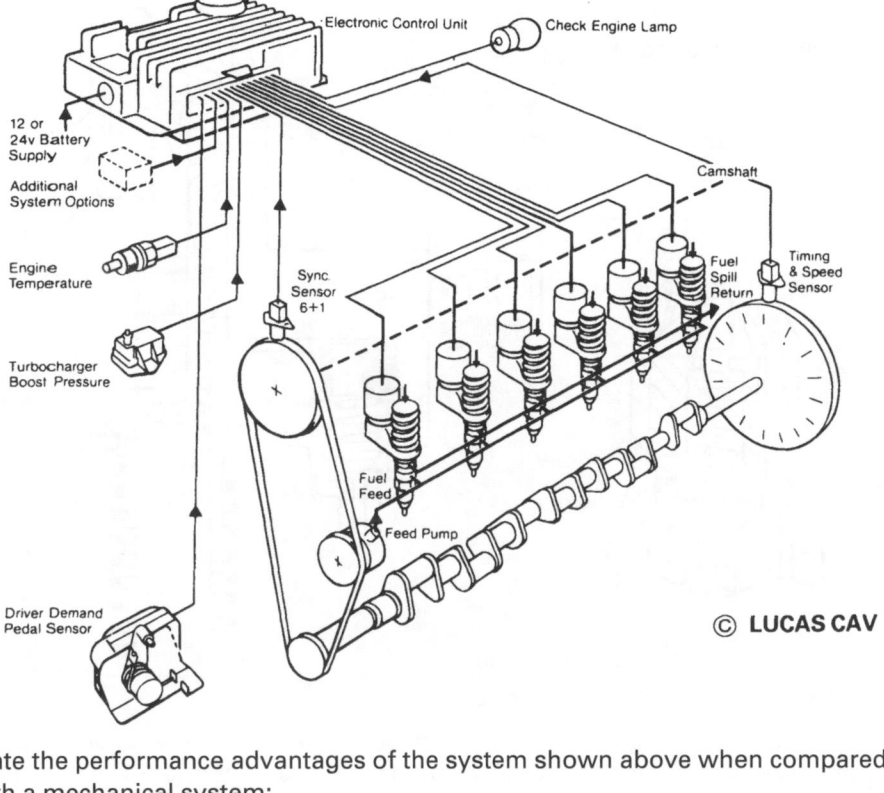

Electronic Control Unit
Check Engine Lamp
12 or 24v Battery Supply
Additional System Options
Engine Temperature
Turbocharger Boost Pressure
Sync. Sensor 6+1
Camshaft
Fuel Spill Return
Timing & Speed Sensor
Fuel Feed
Feed Pump
Driver Demand Pedal Sensor

© **LUCAS CAV**

State the performance advantages of the system shown above when compared with a mechanical system:

....
....
....

Unit Injector

How does the positioning of the injector aid combustion?

....
....
....
....
....
....

SCANIA CYLINDER HEAD SHOWING A SECTION THROUGH TWO CYLINDERS

The injector (shown left) has an 8-hole nozzle tip to assist in the creation of a fine, consistent mist of fuel. Because each injector has an individual pump located virtually at the point of injection, a very high pressure can be maintained at the desired level. This means that the system will not suffer pressure variations that can occur when using complex plumbing from a remote pump.

Why is the engine idling speed much more even with a unit injector system?

....
....
....
....

When is it an advantage to maintain a slow even speed?

....
....
....
....

© **SCANIA**

Name the indicated parts

124

ELECTRONICALLY PROGRAMMED INJECTION CONTROL (EPIC) COMPLETE SYSTEM

The diagram shows layouts of the **Electronic Engine Management System** and **Mechanical Fuel Layout** which includes **Turbo Charger Control** and **Exhaust Gas Recirculation**. It also shows the facility to connect a **Diagnostic Tester (Lucas Laser 2000)** to the electronic system. Note also the water sensor on the fuel filter unit.

Name below the numbered parts shown on the drawing:

1. ....
2. ....
3. ....
4. ....
5. ....
6. ....
7. ....
8. ....
9. ....
10. ....
11. ....
12. ....
13. ....
14. ....
15. ....
16. ....
17. ....
18. ....
19. ....
20. ....
21. ....
22. ....
23. ....
24. ....
25. ....

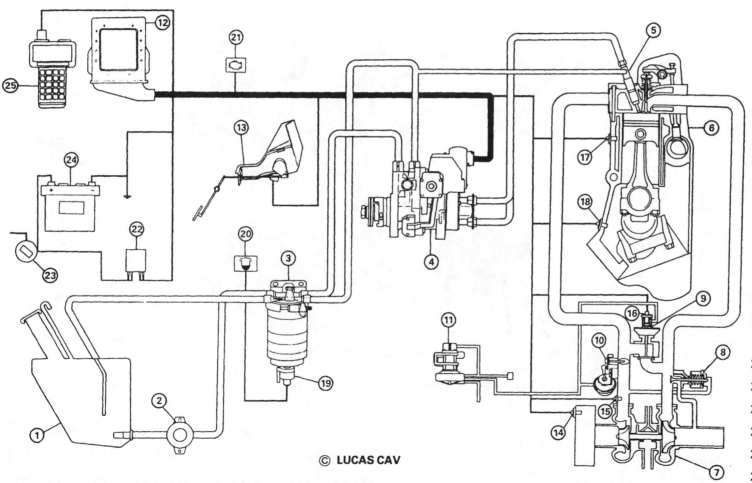

© LUCAS CAV

125

ELECTRONIC FUEL PUMP CONTROL

Lucas Epic 80 Pump for Direct Injection Engines

This electronically programmed injection control (EPIC) pump is a rotary design having many basic mechanical features similar to the DPA pump. There are 4 plungers instead of 2, the cam ring advances in a similar manner, and the rotor and cam rollers and shoes move axially in the pump to increase or decrease the fuel supply. The governor is replaced by solenoids and sensors; these regulate the fuel supply under the electronic control of the ECU.

A schematic drawing of the EPIC 80 Pump–Rotor Control is shown below. Fuel is supplied from the transfer pump (13) via the stop solenoid (4) to the rotor (11). The rotor is moved axially by the fuel supplied from the rotor feed actuator (2); this axial movement allows the shoes to move down the cam slope (12) to reduce fuelling.

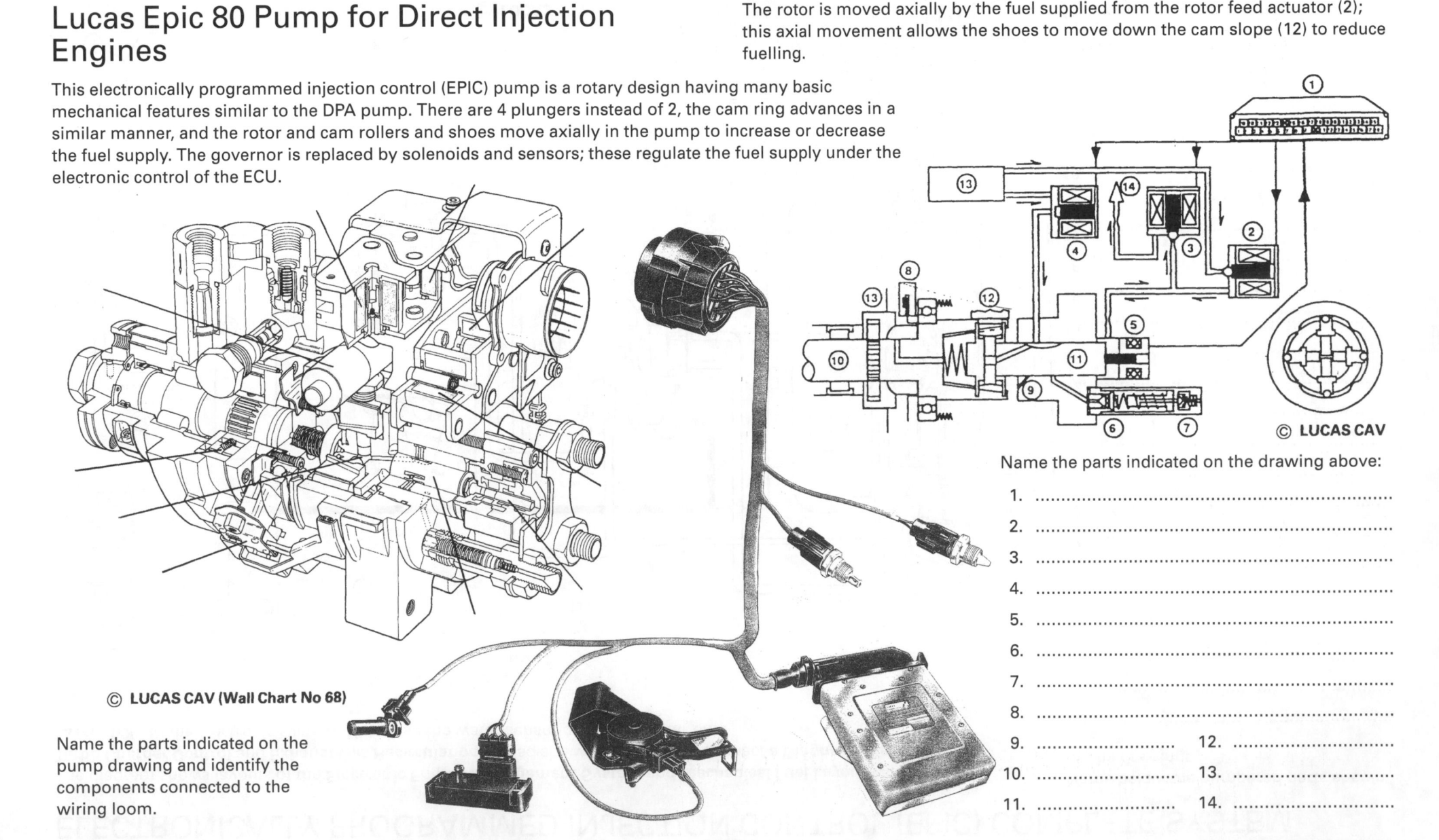

© LUCAS CAV

© LUCAS CAV (Wall Chart No 68)

Name the parts indicated on the pump drawing and identify the components connected to the wiring loom.

Name the parts indicated on the drawing above:

1. ....
2. ....
3. ....
4. ....
5. ....
6. ....
7. ....
8. ....
9. .... 12. ....
10. .... 13. ....
11. .... 14. ....

126

EPIC 80 PUMP – ELECTRONIC OPERATION

Fuel Delivery Control

The rotor feed and drain actuators direct fuel to and from the rotor face to move the drive shaft axially against the pressure of the speed spring. The actuators are closed when energised and no rotor axial movement then occurs. The ECU is therefore continually switching these actuators on and off to regulate the engine speed.

How is the fuel delivery decreased?..

....

....

How is the fuel delivery increased?...

....

....

Name the main parts of the Rotor and Drive assembly.

How is the engine stopped?...

....

....

The EPIC fuel pump has no mechanical linkages. How is the throttle operation transferred to the pump to enable an increase or decrease in fuelling?

....

....

....

COMPONENT ARRANGEMENT OF THE EPIC ENGINE MANAGEMENT SYSTEM (FORD 2.5L DIRECT INJECTION TURBOCHARGED ENGINE)

The ECU takes readings from all the sensors and compares them to pre-programmed three-dimensional maps (see page 256 for examples of similar maps) to determine the optimum injection timings and fuel delivery for the engine at any speed or load.

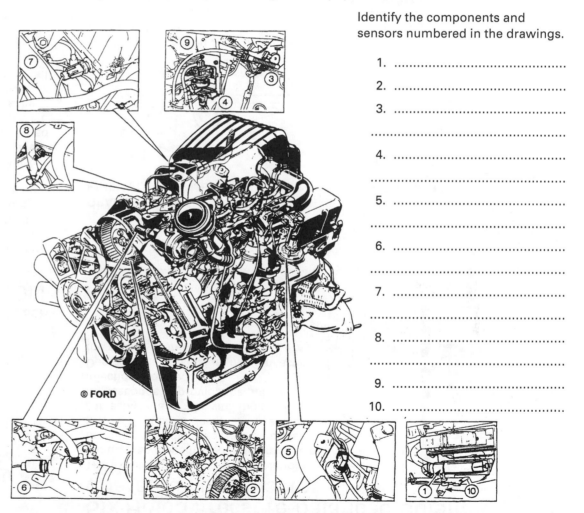

© FORD

Identify the components and sensors numbered in the drawings.

1. ....

2. ....

3. ....

....

4. ....

5. ....

....

6. ....

....

7. ....

....

8. ....

....

9. ....

10. ....

COMMON RAIL INJECTION SYSTEM

The common rail injection system is a high pressure direct injection (HDI) system.
The Bosch system (shown right), employs a high pressure pump which supplies near constant pressure to a large diameter manifold tube or 'rail'. This forms the pressure reservoir, feeding pipes of conventional diameter to the injectors. Electronic operation of the solenoid on each injector gives precise control over the timing and metering of fuel injected into each cylinder.

Name the indicated parts   © BOSCH

FORD – SECOND GENERATION – COMMON RAIL SYSTEM

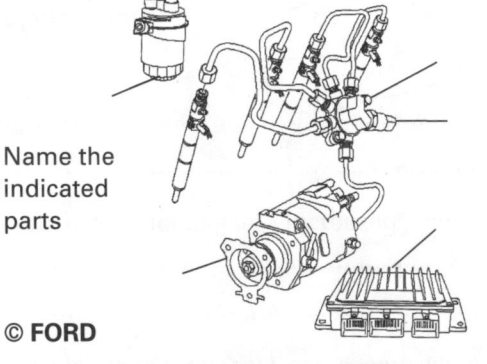

Name the indicated parts

© FORD

On the drawing (shown left) the common rail is a thick walled spherically shaped reservoir, it is fitted with a fuel pressure sensor. The fuel pump is a dual-stage, vane-cell pump that draws fuel from the fuel tank under low pressure and delivers it at very high pressures. Based on operating demands an inlet metering valve measures the precise amount of fuel to be pressurised in the common rail.

ENGINE MANAGEMENT SENSORS

The Injector Driver Module (IDM) controls the common rail system and works in tandem with the Ford EEC-V engine management system. These units obtain sensor information from items such as: fuel metering valve, crankshaft and camshaft position sensors, cylinder head temperature sensor, accelerator, clutch and brake position sensors/switches.
Name SIX other sensor positions used to contribute to the systems operation.

1. ....   4. ....

2. ....   5. ....

3. ....   6. ....

Six Holed Pressure-balanced Solenoid-operated Fuel Injector

The injector contains a small solenoid control valve positioned in the middle of the injector body closer to the nozzle needle than other designs. The solenoid can achieve a rapid switching time of 0.3 milliseconds to provide the pulsed injection required.
Describe the pulsed fuel delivery action of the sectioned injector shown.

....

....

....

....

....

....

Name the indicated parts

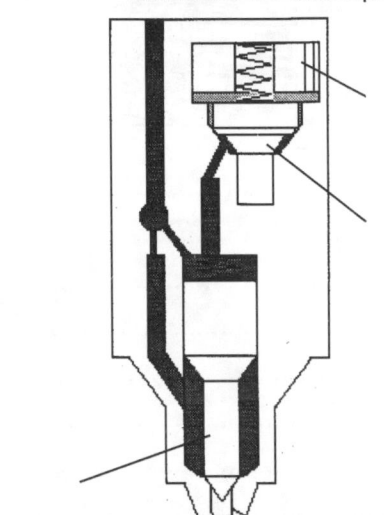

© FORD

GRAPHS SHOW:
WHEN PULSED INJECTION OCCURS

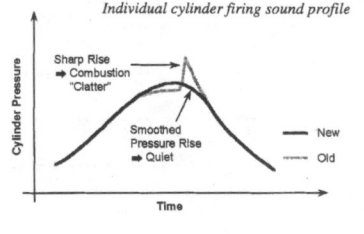

Single cylinder fuel injection profile

What is the effect on combustion caused by the first pulse?

....

....

....

What is the effect of the last pulse shown? Typical post injection is 86°atdc.

....

....

THE PRESSURE RISE IN THE CYLINDER

Individual cylinder firing sound profile

The total effects of the pluses are shown on the graph. Describe what improvements have been made compared to the conventional diesel engine.

....

....

COMMON RAIL SYSTEM – USING AN EXHAUST PARTICULATE FILTER

A particulate filter stops soot and smoke being ejected from the exhaust. It is positioned after the oxidation catalyst inside the catalytic converter box. The filter is made of silicon carbide. The exhaust gas is forced through the porous structure, which collects the contaminates, before the gas leaves the system.

Name the items numbered below

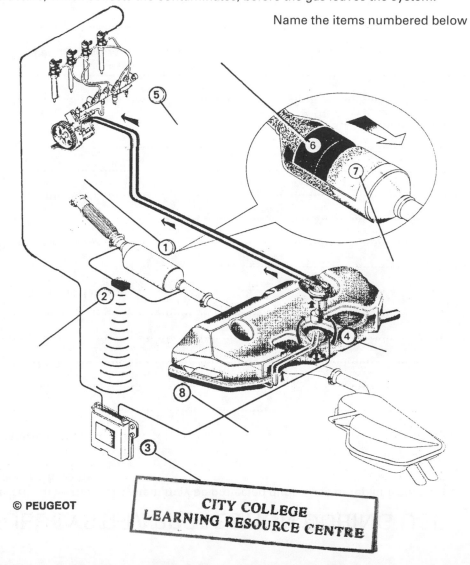

© PEUGEOT

CITY COLLEGE
LEARNING RESOURCE CENTRE

POLLUTION CONTROL

European emission regulations Euro -4 have set a particulate (soot and smoke) limit of 0.025g/km. This low value may be achieved in different ways. The solution taken by Peugeot–Citroën (PSA) is described on this page.

Diesel particulates burn at around 550°C but the exhaust gas in a small diesel engine only reaches 200°C and so cannot naturally get rid of them. The filter must be regenerated every 250 to 300 miles of operation. What is meant by 'regenerated' in this case?

....

Pressure sensors – one either end of the converter – indicate to the engine management system when to start the regeneration – burn – procedure. Describe in TWO stages how the temperature is increased to achieve the burn.

STAGE 1 REDUCE THE NATURAL COMBUSTION TEMPERATURE OF THE PARTICLES

....
....
....
....
....
....
....
....

STAGE 2 INCREASE THE EXHAUST GAS TEMPERATURE

....
....
....
....
....
....
....
....

The complete regeneration (filter cleaning) takes only two to three minutes. What TWO service procedures should be carried out at 50000 miles?

....
....

FUEL SYSTEM PROTECTION DURING USE

Describe how the fuel system may be protected during use or repair from the following hazards:

(a) Corrosion ....
....
....

(b) Freezing ....
....
....

(c) Mechanical damage ....
....

(d) Neglect ....
....

(e) Contamination ....
....
....
....
....

(f) Tampering ....
....
....

(g) Air in system ....
....
....

List the general rules/precautions to be observed while carrying out routine maintenance and running adjustments when:

1. Maintaining cleanliness ....
....

2. Preventing fuel leakage ....

3. Fuel is likely to contact skin ....
....

4. Working on high-pressure fuel lines ....
....
....

5. Avoiding fire hazards ....
....

6. Working in an enclosed space with the engine running ....
....
....

7. Completing the repair job ....
....

8. Road testing ....
....

DIAGNOSTICS: DIESEL FUEL SYSTEM – SYMPTOMS, FAULTS AND CAUSES

State a likely cause for each symptom/system fault listed below. Each cause will suggest any corrective action required.

SYMPTOM/SYSTEM FAULTS	LIKELY CAUSE
Engine revs to above manufacturer's maximum speed, or engine will not reach that speed.	....
When driving along the engine starts to jerk or misfire.	....
Engine makes a louder knocking sound than usual.	....
Engine stops or stalls every time the clutch is disengaged when the vehicle stops.	....
At slow running speed the engine keeps running roughly (hunting).	....
The engine is difficult to start, and requires continual cranking.	....
There is an abnormal metallic rattling noise that keeps occurring.	....
The engine has a distinct lack of power and gives poor acceleration under load.	....
Fuel consumption seems to be excessive and exhaust is smoky.	....
Engine keeps on running when the control key is switched off.	....

DIAGNOSTICS: DIESEL FUEL SYSTEM – Continued

State a likely cause for each symptom/system fault listed below. Each cause will suggest any corrective action required.

SYMPTOM/SYSTEM FAULTS		LIKELY CAUSE
Black smoke is issuing from the exhaust particularly when engine is under load.		....
Blue smoke shows at:	1. any load 2. when starting from rest 3. after changing gear	....
White smoke shows, particularly when engine is cold.		....

List statutory requirements relating to CI fuel systems required by the MOT annual test under the following headings:

(a) Power-to-weight ratio ....
....
....

(b) Smoke emission..
....
....
....

(c) Noise ....
....
....

(d) Fuel and oil leakage ....
....
....

ROUTINE MAINTENANCE

Routine maintenance is carried out on a diesel fuel system to ensure that it maintains its correct supply of fuel at all times. This will maintain the fuel system's efficiency and reliability and so increase its and the engine's working service life.

List TEN typical routine maintenance/service requirements:

1. ....
2. ....
3. ....
4. ....
5. ....
6. ....
7. ....
8. ....
9. ....
10. ....

DIESEL FUEL

This is of the non-volatile type. The approximate chemical composition of a diesel fuel is .... % carbon, ... % hydrogen, plus .... % sulphur and .... % oxygen.

State the safety advantage of using fuel oil as compared with petrol:

....

....

What is the difference between the fuel's 'flash point' and 'self ignition point'?

....

....

....

What is meant by the 'congealing point' of the fuel?

....

....

Ultra-Low-Sulphur Diesel (ULSD)

What is the sulphur content of ultra-low-sulphur diesel (city diesel)? ....
This amount qualifies for a Government duty advantage over conventional DERV. In 1996 the British maximum sulphur limit was 2000 parts per million (ppm) by mass. All low sulphur diesel contains a lubricity additive.

Why is the sulphur content being progressively reduced?

....

CETANE NUMBER

The single most important quality of diesel fuel is its ignition quality. This quality is indicated by its cetane number.
The cetane rating of a fuel is measured by its ....

....

....

How would diesel knock be influenced by the cetane rating?

....

....

The cetane number of a normal diesel fuel is .... It is determined in a somewhat similar manner to that used to determine the octane rating of petrol. However the reference fuels are hexadecane (that is, cetane) and alphamethyl-naphthalene.

ENERGY CONVERSION

The fuel used in a vehicle contains potential heat energy which is stored in a chemical form.

When liberated this heat energy can readily be converted into a mechanical force to propel the vehicle.

The SI unit for heat energy is the ....

Different fuels release differing amounts of heat energy. The Specific Heating Value (calorific value) of a fuel is the quantity of kJ or MJ available to do work when 1 kg of fuel is properly burnt.

Complete the following Specific Heating Value table.

Fuel	Specific Heating Value MJ/kg	Fuel	Specific Heating Value MJ/kg
Petrol Regular		L. P. G.	
Petrol Premium		Alcohol	
Diesel		Paraffin	

Why is it not possible to convert all the heat produced by combustion in the engine into useful work?

....

....

....

The percentage of heat used to do useful work from the fuel used in an engine is approximately .... spark-ignition and ....
compression-ignition.

This percentage is known as the ....

Chapter 8

Battery, Charging and Starter Systems

Lead–acid battery construction	135	Checking for faults – external components	150
Battery charging	136	Diagnostics	151
Battery states of charge and capacity	137	Types of starter motor	152
Connecting batteries in series or parallel	138	Electric motor wiring	152
Battery maintenance	138	Motor operation	153
Battery testing	139	Basic components in a starter motor	154
Maintenance-free battery	140	Construction of a solenoid	154
Diagnostics	140	Starter motor construction	155
Battery jump starting	141	Inertia drive	155
Vehicle memory saver	141	Lucas M35J pre-engaged starter motor	156
Safe working practice	141	Drive mechanism – pre-engaged starters	157
Battery charging systems	142	Permanent magnet, gear reduction starter motor	158
Functions of semiconductor components	142	Axial starters	159
The alternator	143	Coaxial starters	160
Full-wave static rectification	144	Testing starter motors	161
Lucas ACR type alternator	145	Inspection for faults	163
Typical Lucas 127 range alternator	146	Diagnostics	164
Electronic regulator control	147	System protection during use	165
Alternator performance check	148	Good working practice	165
Alternator removal and repair	149		

LEAD–ACID BATTERY CONSTRUCTION

The lead–acid battery is used on most automobiles in either 12 or 24-V form.

A 12-V battery container consists of six separate compartments. Each compartment contains a set of positive and negative plates; each set is fixed to a bar which rises to form the positive or negative terminal.

Between each plate is an insulating separator.

Name the parts on the sketch below.

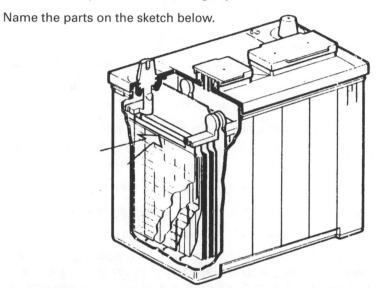

Since each cell has a nominal electrical pressure of 2 V, to produce a 12-V battery six cells must be joined together in series.

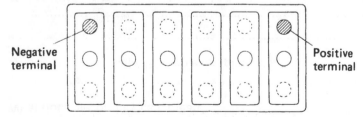

Negative terminal

Positive terminal

The cells are connected by buss bars (or links), not normally seen on modern batteries. Show where these would be connected in the diagram above and indicate the polarity of each cell by symbols.

The battery main components are shown below. Name the parts and describe their construction and function.

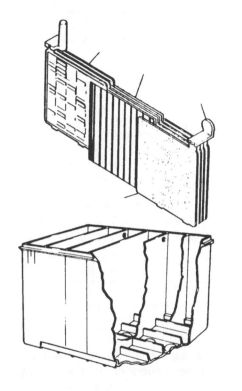

Name the types of battery cable connectors shown below.

 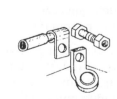 

....

PLATE GRIDS

....

....

....

....

....

....

....

....

....

....

....

....

....

SEPARATORS

....

....

....

....

....

....

CONTAINER (or CASE)

....

....

....

....

....

BATTERY CHARGING

The battery is a series of chemical cells. Each cell of a lead–acid battery is capable of producing 2 V. The size and number of plates in the cell determine its capacity or output. It is of a secondary cell type.

What does 'secondary cell' mean?

....

What is a primary cell? Give an example:

....

....

The positive and negative plate active materials of the secondary cell, although of a similar base, have a different chemical composition and when submerged in a suitable electrolyte produce an electrical pressure difference which when connected to a circuit allows a current to flow. The chemical reaction then discharges the cell until the plates become chemically similar.

Why is the cell then capable of being recharged?

....

....

....

When the battery requires charging, the supply, be it from the vehicle's charging system or the mains supply, must be a direct current (dc). Why must only direct current be used?

....

....

Complete the table to show how the plate materials and electrolyte are affected during the charge and discharge cycle.

Process	Positive plate	Electrolyte	Negative plate
Fully charged			
Discharged			

During the charge process the chemical reaction causes the plates to give off bubbles. Name the gas given off at the:

positive plates ....   negative plates ....

When batteries are charged on a vehicle by the alternator, they are initially (after engine starting) subjected to a high rate of charge and then controlled to a much lower charge rate.

When removed from the vehicle and externally charged, fast or slow chargers may be used.

It is preferable to .... charge if time allows.

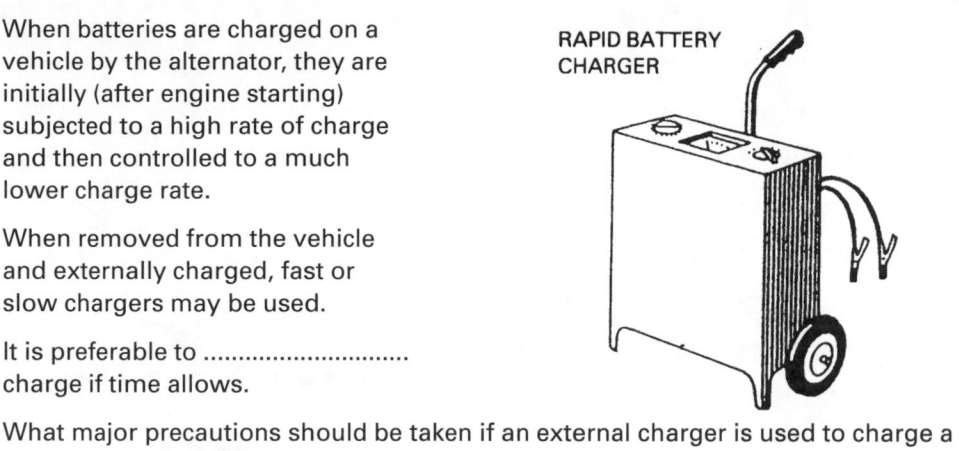

RAPID BATTERY CHARGER

What major precautions should be taken if an external charger is used to charge a battery still connected to the vehicle?

....

....

....

....

Show on the sketch how three 12-V batteries are connected to a constant current charger.

Constant current battery charger

The cells are electrically connected in:

....

When connecting a battery to an external charging system, what way (electrically) are the charger-to-battery connections made?

....

....

List the precautions required when charging batteries:

....

....

....

....

BATTERY STATES OF CHARGE AND CAPACITY

The performance of a battery is affected by many factors; these include its state of charge, temperature, condition and age.

Complete the graph below to show the variation of internal electrical reistance throughout the charge cycle. When is the resistance at its lowest and what would cause it to increase?

....
....
....
....
....
....
....

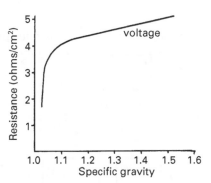

COLD CRANKING PERFORMANCE

Batteries are given a rating to indicate their ability to crank the engine at – 18°C. At this or lower temperatures the capacity of the battery is greatly reduced.

Complete the graph below to show how the capacities of a fully charged and 80% charged battery are reduced by a fall in temperature to – 18°C.

The values show the amount of current being drawn from the battery. Describe the effects of low temperature.

....
....
....
....
....
....

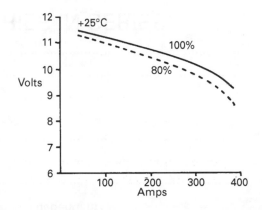

Why should batteries never be completely discharged?

....
....
....
....
....
....

Show by means of a graph the effect of the charging cycle on relative density (specific gravity) and cell voltage:

....
....
....
....
....
....

State effects caused by excessive overcharging:

....
....
....

State effects caused by lack of charging or battery standing idle:

....
....
....
....

On what is the capacity of a battery primarily dependent?

....

....

....

What can reduce the capacity of a battery?

....

....

....

CONNECTING BATTERIES IN SERIES OR PARALLEL

The capacity or output voltage of a system can be changed by connecting batteries together.

Show the pairs of batteries connected in:

(a) series   (b) parallel

What will be the effect on voltage output and capacity if batteries are connected in:

1. series? ....

....

....

2. parallel? ....

BATTERY MAINTENANCE

What maintenance checks are indicated on the drawing opposite?

1. ....

....

2. ....

....

3. ....

....

4. ....

....

5. ....

....

© LUCAS

Which of the above items would not require checking on a modern 'Low maintenance' or 'Maintenance free' battery? Give reasons.

....

....

....

If a battery is suspected of being faulty, a systematic check of its condition should be carried out.

State the basic visual and manual checks that should be made:

1. ....

2. ....

3. ....

4. ....

When testing the battery, equipment such as voltmeter, hydrometer, high rate discharge tester and battery charger may be used. List typical faults that testing may reveal:

1. ....   4. ....

2. ....   5. ....

3. ....   6. ....

BATTERY TESTING

Carry out tests 1, 2 and 3 on a battery, and state the function or make comments on the SIX tests mentioned on this page.

1. OPEN CIRCUIT VOLTAGE

Switch on headlamps for 30 seconds, then switch off all electrical loads. Connect voltmeter across battery terminals to determine the stabilised open circuit voltage. Expected voltage readings are:

....

....

Reading obtained .... Comment ....

This test would replace the hydrometer test on sealed batteries.

2. HYDROMETER TEST

How does this test indicate the condition and state of charge of the battery?

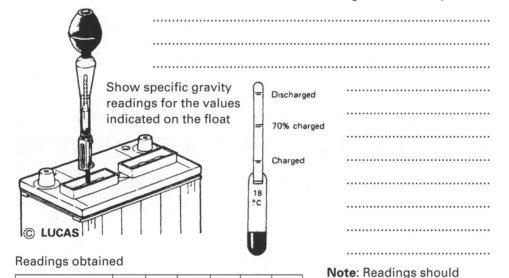

Show specific gravity readings for the values indicated on the float

© LUCAS

....

....

....

....

....

....

....

....

....

Readings obtained

Cell number	1	2	3	4	5	6
Sp. gr. value						

Note: Readings should indicate over 70% charged before carrying out further tests.

Comment...

3. HIGH-RATE DISCHARGE TEST (Battery load test)

This is made to ensure that each cell will supply the heavy currents required for starting.

....

....

....

....

....

....

© LUCAS

Actual reading ....

Expected reading ....

Comment ....

Note: Batteries that fail the hydrometer and battery load test should be replaced.

4. CYCLING TEST ....

....

....

....

....

5. CAPACITY TESTS   Reserve capacity ....

....

....

....

10/20 hour rate ....

....

....

6. COLD START PERFORMANCE ....

....

....

....

MAINTENANCE-FREE BATTERY

These batteries are completely sealed so a hydrometer state of charge cannot be made. How then is the state of charge determined?

....
....
....
....
....

This type of battery is fitted with lead–calcium plate grids to reduce gassing to a minimum. The gassing that occurs when fully charged is passed into a gas reservoir in the lid and returned as it forms water droplets.

How is gassing controlled during the charge process?

....
....
....
....
....

DIAGNOSTICS: BATTERY – SYMPTOMS, FAULTS AND CAUSES

State a likely cause for each symptom/system fault listed below. Each cause will suggest any corrective action required

SYMPTOM	SYSTEM FAULT	LIKELY CAUSE
Electrical systems are completely dead or lamps are very dim and starter fails to operate.	No or low terminal voltage	....
Starter will not crank engine first thing in the morning, but is OK after starting.	Loss of capacity and failure to hold charge	....
Starter cranks engine slowly, alternator seems to be charging correctly.	Incorrect relative density of electrolyte or acid loss	....
Starter will not crank engine – there is a wet pool on the floor.	Physical damage	....
Slow cranking of engine. Battery needs continual topping up.	Undercharging Overcharging	....
Batteries do not become charged in the expected time – battery gasses excessively	Faulty operation of multi-battery layouts	....

BATTERY JUMP STARTING

State the connecting procedure using a slave battery.

....

....

....

....

....

....

....

....

Indicate on the drawing by placing numbers in the circles in the correct order, how to connect a slave battery to a vehicle using jump leads.

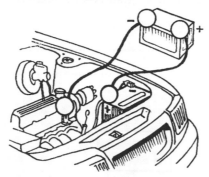

Battery Jump Starters

© GUNSON

It is much more convenient and safe to use a portable re-chargeable battery pack to start a vehicle having a discharged battery, than to use jump leads and a spare battery or a battery still fixed to another vehicle.

These battery packs provide a 12 V power source fitted in a sturdy carrying case. The lead acid battery used is a sealed (gel-electrolyte) type designed for cyclic use and can operate at a maximum boost of up to 400 A.

The unit's basic use is to act as a 'booster' to start a vehicle on a cold morning.

What other features may these units possess?

....

....

....

....

How should a Jump Starter Unit be kept in good condition?

....

....

VEHICLE MEMORY SAVER

© GUNSON

On modern vehicles there are many memory controlled devices which would need to be reset should the battery be disconnected. To prevent this inconvenience a memory saver may be used.

What vehicle units may suffer from loss of memory due to battery disconnection?

....

....

....

....

The vehicle memory saver is a unit which provides a 12 V supply to the electrical system when the battery is disconnected. The supply may be obtained from the mains via a transformer if in the workshop or by a 12 V battery if on breakdown. Where is the memory saver connected to the vehicle's electrical system?

....

SAFE WORKING PRACTICE

State general rules and precautions to be observed before working on batteries and charging systems relative to the following:

Handling batteries ....

....

Charging batteries ....

....

....

Connecting batteries or alternators ....

....

....

BATTERY CHARGING SYSTEMS

The type of charging system found on all modern vehicles consists of an alternator and storage battery. On many older vehicles (such as Volkswagen Beetle and Land Rover) a dynamo is used to produce electric current.

State the purpose and functional requirements of (a) the charging system and (b) the battery:

(a)...

(b) ....

....

....

Complete the diagram to show a typical layout of a battery-charging system. Name the parts.

Alternator complete with voltage regulator control

....

....

....

....

....

....

....

FUNCTIONS OF SEMICONDUCTOR COMPONENTS

The electronic components listed below are all used in the alternator's electrical circuit and may be used in any other circuit that adopts electronic control.

The alternating current flow must be rectified to direct current. This is achieved by using a number of static rectifiers called diodes.

The function of a diode is to: Show a diode's electrical symbol and indicate the direction of current flow.

....

....

It is made from ....

How is this material made into a semi-conductor?...

....

....

....

Show the electrical circuit symbol for components below and explain their basic function. Show directions of current flow.

Transistor   Avalanche diode

....

....

Surge protection diode

....

....

....

....

THE ALTERNATOR

In order to produce an electric current by magnetic induction, three basic requirements must be fulfilled.

In an alternator these are:

....

....

....

....

....

....

....

....

....

....

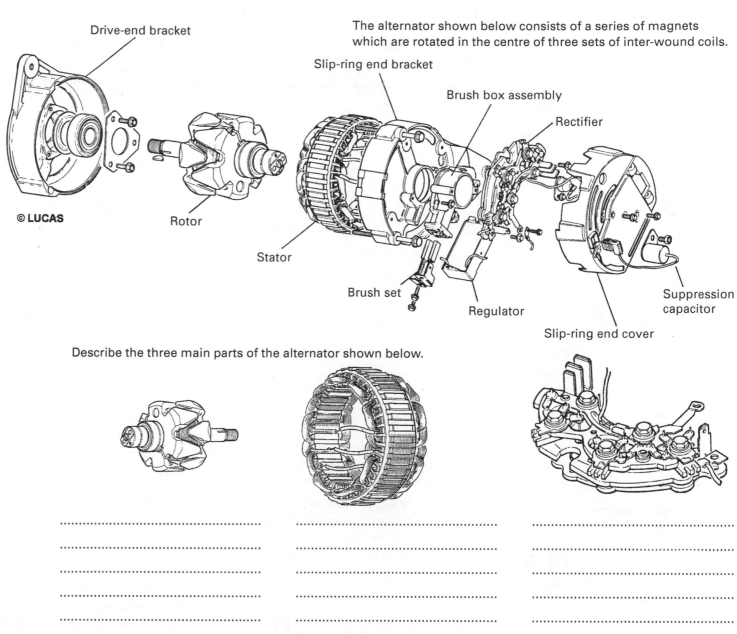

Drive-end bracket

© LUCAS

Rotor

Stator

Brush set

The alternator shown below consists of a series of magnets which are rotated in the centre of three sets of inter-wound coils.

Slip-ring end bracket

Brush box assembly

Rectifier

Regulator

Slip-ring end cover

Suppression capacitor

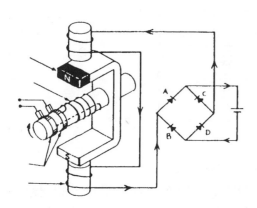

Label the basic requirements on the simple sketch above.

Describe the three main parts of the alternator shown below.

....

....

....

....

....

....

FULL-WAVE STATIC RECTIFICATION

When the magnet in a simple alternator is revolved, one complete turn, an emf is induced in the circuit, first in one direction and then in the reverse direction.

On the diagram below show an alternating pulse and a rectified pulse.

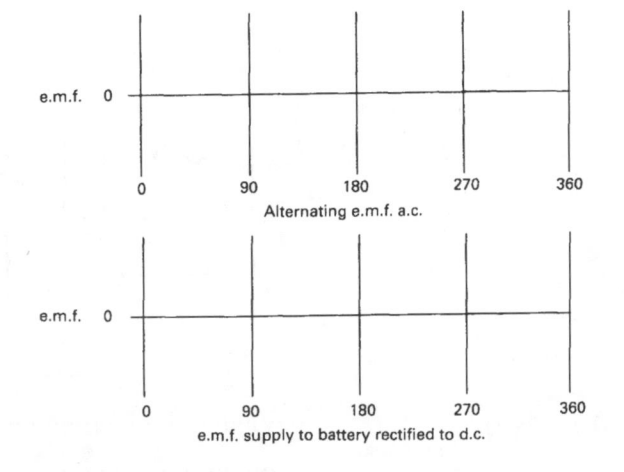

Alternating e.m.f. a.c.

e.m.f. supply to battery rectified to d.c.

The components that allow this rectification to occur are diodes.

SINGLE-PHASE RECTIFICATION

The four diodes shown make up a single full-wave bridge rectification system.

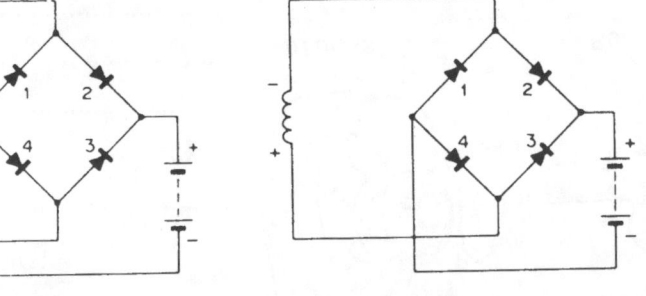

Show, using arrows, how diodes rectify the supply of current induced in a single coil of wire.

Diodes passing current nos ....   Diodes passing current nos ....

THREE-PHASE RECTIFICATION

Name the parts indicated on the rotor shown.

One magnet segment is indicated on the two diagrams below. As there are six magnetic segments (claws) on each magnet there will be twelve voltage pulses induced into each winding every rotor revolution.

Trace the rectified flow path through the system on each diagram and describe the action.

....
....
....
....
....
....
....
....
....
....
....
....
....
....
....
....

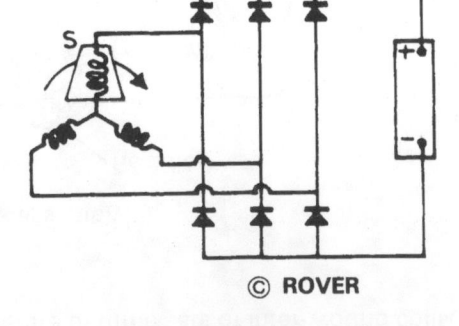

© **ROVER**

LUCAS ACR TYPE ALTERNATOR

Name the major parts of the alternator shown below.

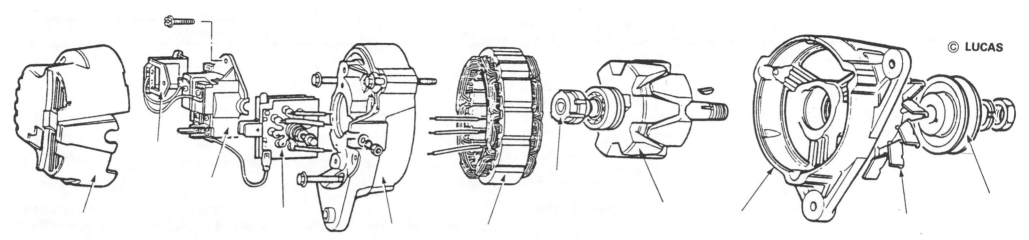

© **LUCAS**

A wiring diagram is shown below for the alternator. Identify the items indicated and compare them with the physical parts above.

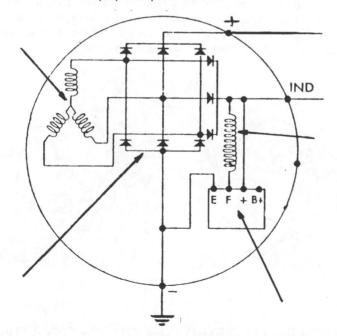

Why is the voltage regulator needed to control output?

....

....

....

How does the voltage regulator control output?

....

....

....

....

....

Why is the output current said to be self-regulating?

....

....

....

....

TYPICAL LUCAS 127 RANGE ALTERNATOR

Name the parts indicated.

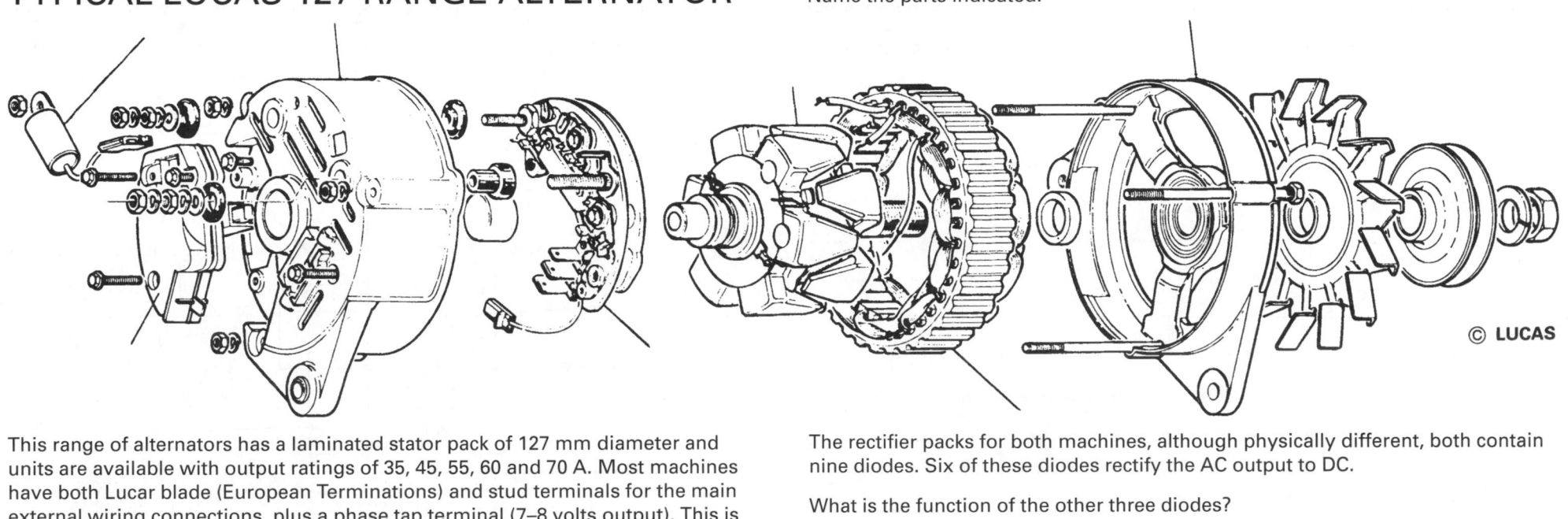

© LUCAS

This range of alternators has a laminated stator pack of 127 mm diameter and units are available with output ratings of 35, 45, 55, 60 and 70 A. Most machines have both Lucar blade (European Terminations) and stud terminals for the main external wiring connections, plus a phase tap terminal (7–8 volts output). This is used on some vehicle applications for a carburettor choke control or operating a tachometer.

Compare this machine with the ACR type shown on the previous page and state how the following components differ:

Regulator assembly...

....

....

Rectifier pack...

....

....

Rotor slip rings...

....

....

The rectifier packs for both machines, although physically different, both contain nine diodes. Six of these diodes rectify the AC output to DC.

What is the function of the other three diodes?

....

....

What is the function of all nine diodes when the engine is stopped?

....

....

What do models using external regulators usually require to prevent battery discharge through the rotor (field) windings when the engine is stopped?

....

....

On a diesel-engined vehicle what other component may be driven by an extension of the centre shaft of the alternator?

ELECTRONIC REGULATOR CONTROL

Most alternators fitted on UK vehicles use electronic regulators fitted at the rear-end of the alternator. This makes the unit a complete self-contained electrical power source.

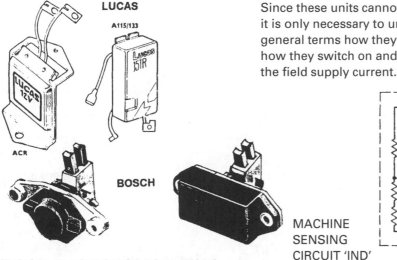

LUCAS
A115/133

ACR

BOSCH

TYPICAL ELECTRONIC REGULATORS

Since these units cannot be adjusted it is only necessary to understand in general terms how they work, that is, how they switch on and then control the field supply current.

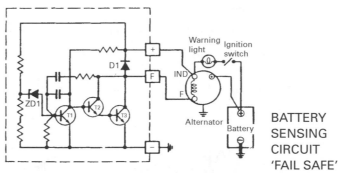

MACHINE SENSING CIRCUIT 'IND'

Examine the machine sensing regulator drawing.

When the ignition is switched on, current flows through the warning light to the IND connection on the alternator, then to + on regulator, through resistor to switch T1 on. This allows flow through the alternator field coil to F. T2 switches T3 on and the current flows to earth unrestricted.

Describe the alternator (field) control as the engine commences running:

....

....

....

....

....

....

....

Alternators may be 'machine-sensed' or 'battery-sensed'.

The term sensing relates to where the regulator picks up the supply current flowing through the regulator to the zener diode.

If machine-sensed the supply is via the three field diodes, through the field winding, to the regulator.

If battery-sensed, the regulator sensing circuit is connected directly by a separate cable to the battery. With this arrangement the alternator is more sensitive to the charging loads placed upon the battery.

© LUCAS

BATTERY SENSING CIRCUIT 'FAIL SAFE'

What is the function of the surge protection diode which is an integral part of the voltage regulator shown above?

....

....

....

On earlier type alternators the surge protection diode is positioned on the alternator's end plate next to the regulator. The connections electrically are in the same position – between IND and earth.

What other methods offer protection to the alternator?

....

....

....

....

147

ALTERNATOR PERFORMANCE CHECK

If an alternator is suspected of being faulty it should be tested on the vehicle before being dismantled.

After completing the simple preliminary checks, carry out the basic alternator tests shown below.

Describe the procedures and show where the meters and leads should be fitted.

List preliminary checks that should be made before alternator is tested:

....

....

....

....

Test 1. Alternator output

Expected output ....

Actual reading ....

....

....

....

....

....

....

....

....

Test 2. Charging circuit

V_1 actual reading ....

V_2 actual reading ....

....

....

....

....

....

....

....

....

Test 3. Check alternator control

Actual reading ....

....

....

....

....

....

....

....

....

ALTERNATOR REMOVAL AND REPAIR

Describe the procedure to remove an alternator from an engine:

....

....

....

....

Describe important points to observe when dismantling and reassembling an alternator:

....

....

....

....

....

....

List visual checks that can be made when the alternator is dismantled:

1. ....

2. ....

3 ....

....

4. ....

5. ....

....

6. ....

7. ....

8. ....

State the basic tests that are being made to the alternator components shown. Carry out similar tests on dismantled alternators.

....

....

....

© LUCAS

....

State another check that could be made to some alternators:

....

....

149

CHECKING FOR FAULTS – EXTERNAL COMPONENTS

The modern electronically controlled alternator has no external components except the ignition warning light and the connections, switches, relays and fuse links to the battery. How should these be checked if a fault is evident?

....

....

....

....

Systems using external regulators may employ field cutout relays and warning light control units; these should be checked for correct operation.

Identify the components:

 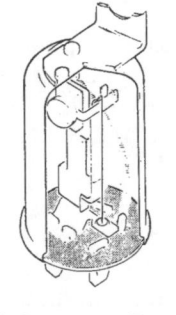

© **LUCAS**

What is the function of the warning light control?

....

....

....

Why is such a control unit not fitted on an alternator with internal regulator control?

....

....

....

....

What mechanical faults occur to electromagnetic regulators?

....

....

List typical checks that should be made on mechanical regulators:

1. ....

2. ....

3. ....

....

4. ....

State considerations given before fitting a new or exchange:

(a) control unit ....

....

(b) alternator ....

....

PROTECTION OF ALTERNATOR FROM ELECTRICAL MISUSE

The electronic components in an alternator have a long life if not subjected to certain hazards during use or repair.

The two most common hazards are:

1. ....

....

2. ....

....

What items protect against these faults?

....

....

DIAGNOSTICS: ALTERNATOR SYSTEM – SYMPTOMS, FAULTS AND CAUSES

State a likely cause for each symptom/system fault listed below. Each cause will suggest any corrective action required.

SYMPTOM	SYSTEM FAULT	LIKELY CAUSE
Battery cranks engine slowly, it is in a low state of charge or electrolyte is low.	No low or high current output	....
Alternator is making an unusual noise and is too hot to touch.	Overheating	....
A squealing noise when engine is revved. A whining noise.	Abnormal noises	....
Light remains on or is intermittent — light not operating before starting.	Incorrect warning lamp operation	....

Diagnostic Test Equipment

List the test equipment used and available for testing batteries and alternators:

1. ....   6. ....

2. ....   7. ....

3. ....   8. ....

4. ....   9. ....

5. ....   10. ....

Use of test equipment:

....

....

....

....

Modifying System

State reasons why a system may be uprated or modified by:

1. Fitting a larger capacity battery and alternator with higher output

....

....

....

....

....

2. Fitting a multi-battery layout ....

....

....

TYPES OF STARTER MOTOR

A starter motor has one basic functional requirement:

....

....

The different sizes and types of engines make it necessary to require different designs of starter motor.

Name the types of drive shown below.

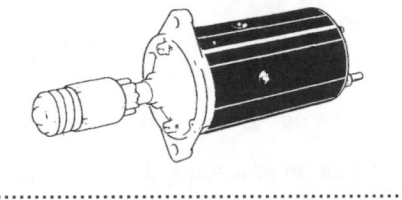

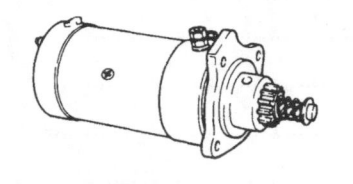

....

....

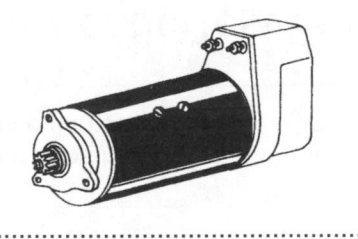

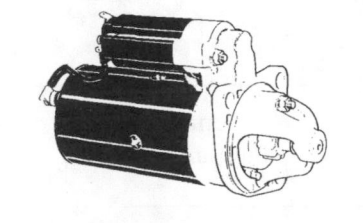

....

....

The components below are used in a pre-engaged starter system. Name the parts and complete the drawing to show a pictorial wiring diagram.

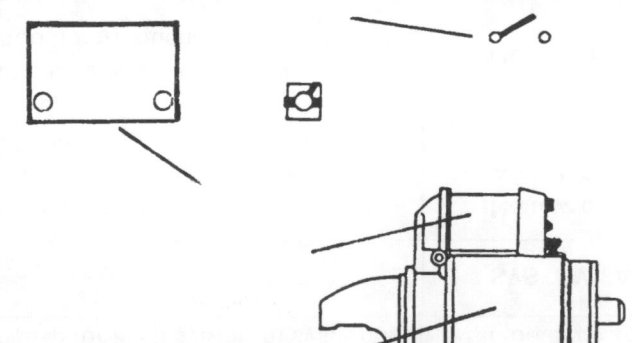

ELECTRIC MOTOR WIRING

Field windings can be wired in a number of ways. Complete the diagrams to show the basic theoretical methods listed below.

Series motor

Shunt motor

Compound motor

What factors determine the type of design used?

....

....

....

....

State the characteristics possessed by a:

Series wound starter motor ....

....

....

....

....

Parallel (shunt) wound general purpose motor ....

....

....

....

....

Compound motor ....

....

....

....

....

MOTOR OPERATION

Describe, with the aid of the simple diagrams below, the basic operation of an electric motor (such as a starter motor):

Armature winding

....
....
....
....
....
....
....
....
....
....

Windings increased

....
....
....
....
....
....
....
....
....
....
....

Field windings connected in parallel

Examine a starter armature and count the number of segments on an actual commutator.

No. of SEGMENTS ....

No. of LOOPS ....

The basic electrical principles of all starter motors are the same and can be demonstrated by passing a current through a coil of wire positioned between the poles of a permanent magnet and noting its effect.

Show and explain this action with the aid of the simple diagrams below.

Complete the second diagram.

No current flowing

Current flowing

....
....
....
....
....
....
....
....
....
....

In practice, most starter motors use four pole-shoes surrounded by field coils and so produce double the magnetic effect. Describe the current flow through the starter shown.

....
....
....
....
....
....
....
....
....
....

MOVEMENT
CURRENT FLOW
MAGNETIC FIELD

BASIC COMPONENTS IN A STARTER MOTOR

A light vehicle starter motor will consist of the following components: armature, commutator end bracket, brush assembly, field magnets (electro or permanent), outer casing (yoke) and gear drive assembly.

Name the parts indicated on the drawings and state their basic functions:

Field windings

....

....

....

....

....

....

© **LUCAS**

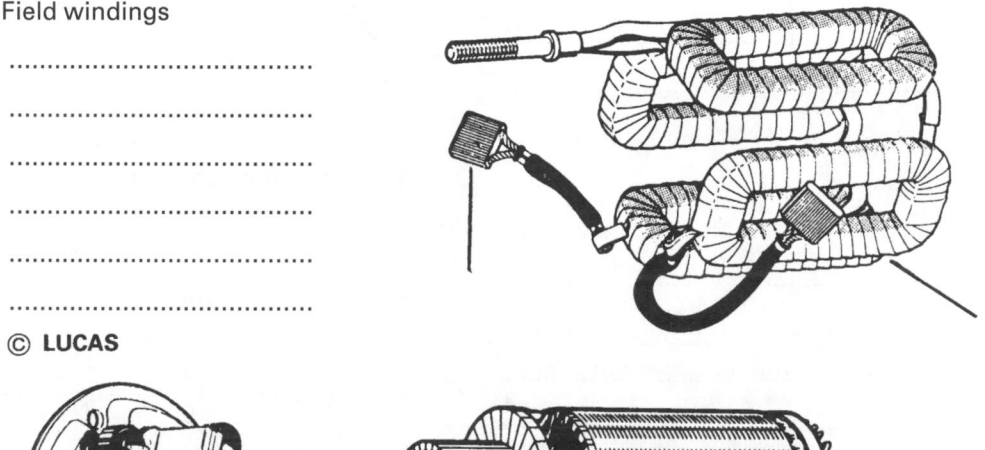

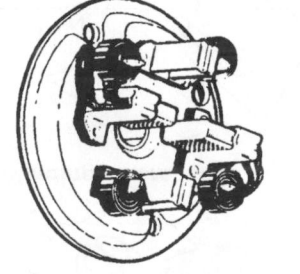

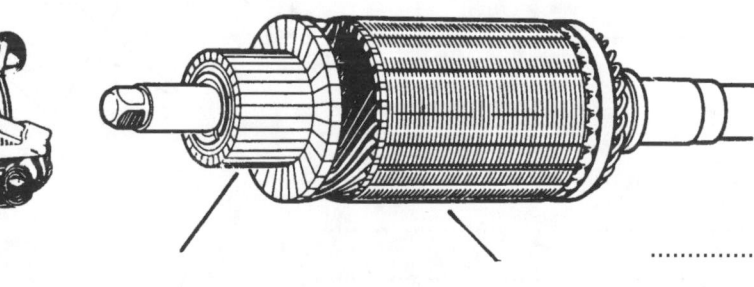

Brush gear ....

Yoke ....

Bearings ....

State the materials used for the following:

field windings ....

armature windings ....

brushes ....

bearings ....

Armature

....

....

....

....

....

....

CONSTRUCTION OF A SOLENOID

The solenoid used for starter motor operation is a heavy-duty type of electromagnetic switch. With the aid of the sectioned sketch below describe the basic construction and operation of a solenoid switch.

....

....

....

....

....

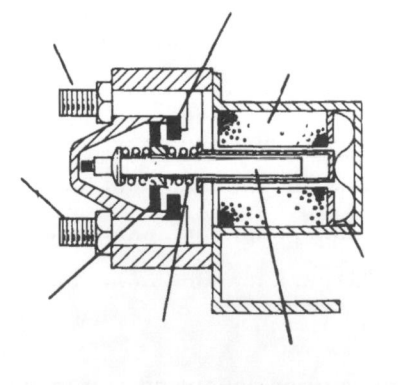

154

STARTER MOTOR CONSTRUCTION

The starter motor shown opposite is of a conventional series parallel construction.

Trace the path of current flow through the motor by stating each component through which it passes.

Item	Current flow
1	*Feed terminal*
2	*Field coils*
3	
4	
5	
6	

Modern starters are mainly wound in some form of series pattern, that is, using a heavy strip field winding in series with the armature so that the field and armature current are equal. What amount of current is required and when does the starter achieve its maximum torque?

....
....
....
....
....
....

Mark the diagram with the item numbers.

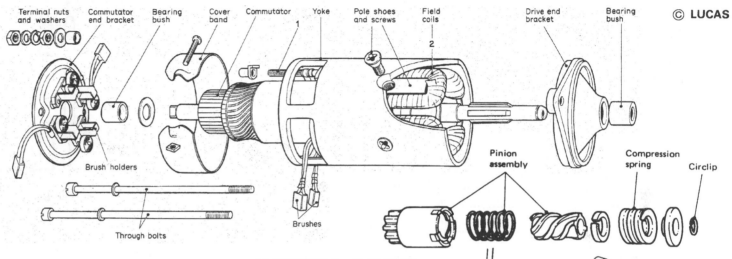

Terminal nuts and washers — Commutator end bracket — Bearing bush — Cover band — Commutator — 1 — Yoke — Pole shoes and screws — Field coils — 2 — Drive end bracket — Bearing bush — © LUCAS

Brush holders

Through bolts

Brushes

Pinion assembly — Compression spring — Circlip

Motor armature

Show the wiring layout for the above starter motor and indicate the flow path through the circuit.

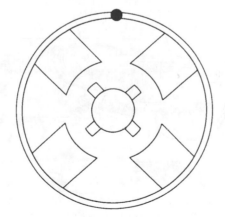

Series parallel

INERTIA DRIVE

As shown, the drive consists of a pinion, mounted on the armature shaft.

Describe, with the aid of the drawings, its operation when the starter turns:

....
....
....
....
....
....
....

155

LUCAS M35J PRE-ENGAGED STARTER MOTOR

Name the parts indicated.

© **LUCAS**

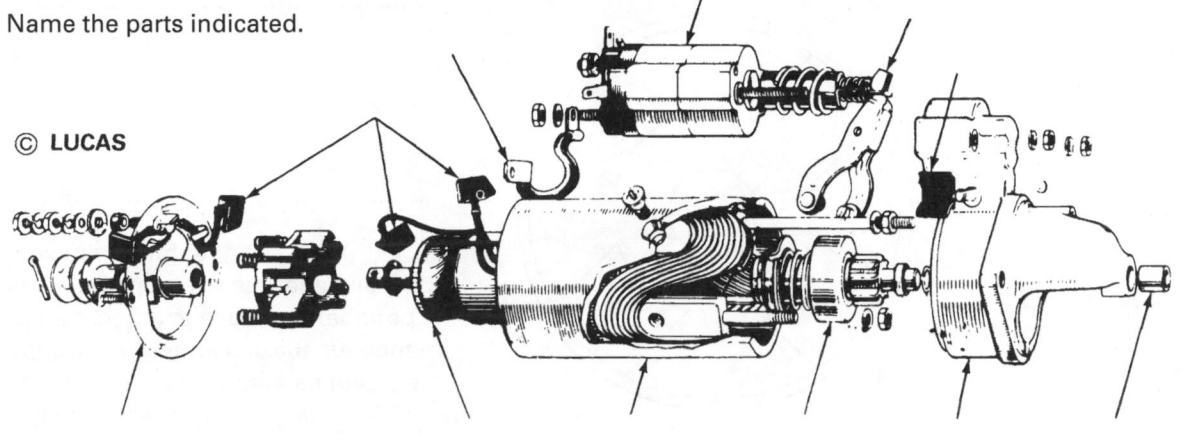

The diagram shows a more modern design of starter motor than the one shown on the previous page.

State how the components named below differ:

Field windings...

....

....

Commutator...

....

Brushes..

....

Brush carrier ....

....

Name the parts on the wiring diagram and indicate the electrical flow through the circuit.

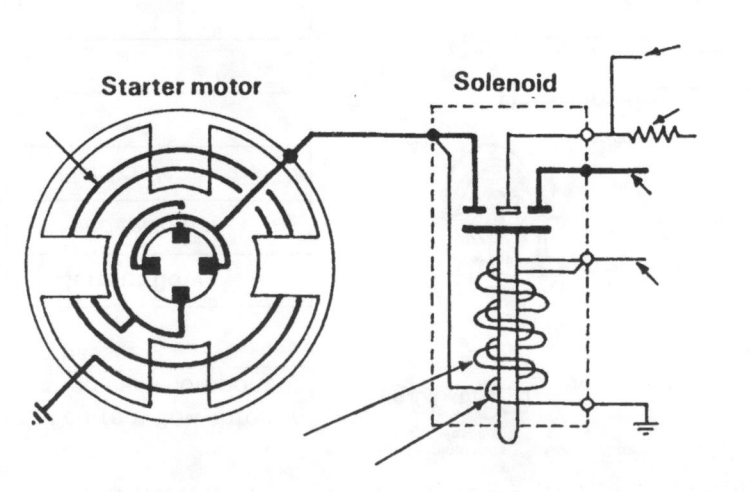

Describe the operation of the starter shown:

....

....

....

....

....

....

....

....

....

....

....

....

....

....

DRIVE MECHANISM – PRE-ENGAGED STARTERS

Movement of the solenoid plunger results in a similar movement of the pinion gear, but in the opposite direction. This allows the gear to engage the flywheel before the armature rotates. What is the purpose of the engagement spring on the pinion?

PIVOT LINKAGE (Name the main parts)

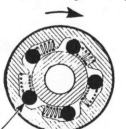

....

....

....

....

....

....

....

....

ROLLER CLUTCH ASSEMBLY

As shown, the rollers are spring loaded and according to direction of drive are either free running or wedge locked.

What is the object of having this unit in the drive?

....

....

....

© **LUCAS CAV**

What are the advantages of the pre-engaged starter motor when compared with the inertia type?

....

....

....

....

PLATE CLUTCH DRIVE

When the starter is required to transmit heavy torques, a multi-plate clutch drive is often used (as shown below).

Describe how the parts are assembled together:

....

....

....

....

Explain how this type holds and disengages drive, and protects against overspeed and overload:

....

....

....

....

....

....

....

....

....

Rivet Pinion retaining ring Barrel unit Thrust washer Backing ring Clutch plates inner outer Helical splines Driving sleeve Circlip Lock ring

Pinion Helical splined sleeve Cushion spring Ring nut Pressure plates Shim Moving member Retaining washer Engagement bush

ARMATURE BRAKING

Some pre-engaged starter motors are fitted with an armature brake assembly. This is located in the armature end bracket as shown, how does it operate?

....

....

....

....

Name
the
parts
indicated.

© LUCAS

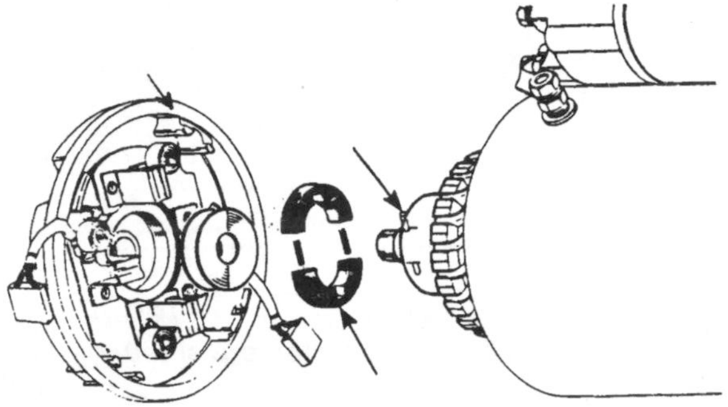

What are the two basic functions of the brake assembly?

1. ....

2. ....

....

Name
the
parts
indicated.

© LUCAS

PERMANENT MAGNET, GEAR REDUCTION STARTER MOTOR

The Lucas M78R/M80R starter motor shown below has a much simpler electrical design than the starters already described. What feature makes it electrically simpler?

....

....

....

The armature of this starter motor rotates at a higher speed than the conventional type and incorporates an epicyclic reduction gear assembly to increase the torque output of the motor.

Describe the operation of the epicyclic gear drive:

....

....

....

....

....

....

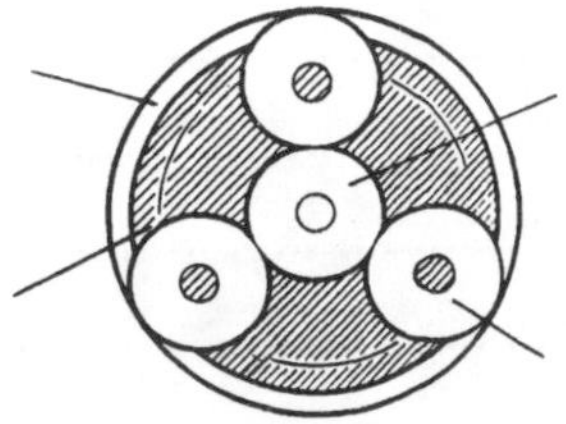

Name the gear drive parts and indicate the rotational movement of the gears.

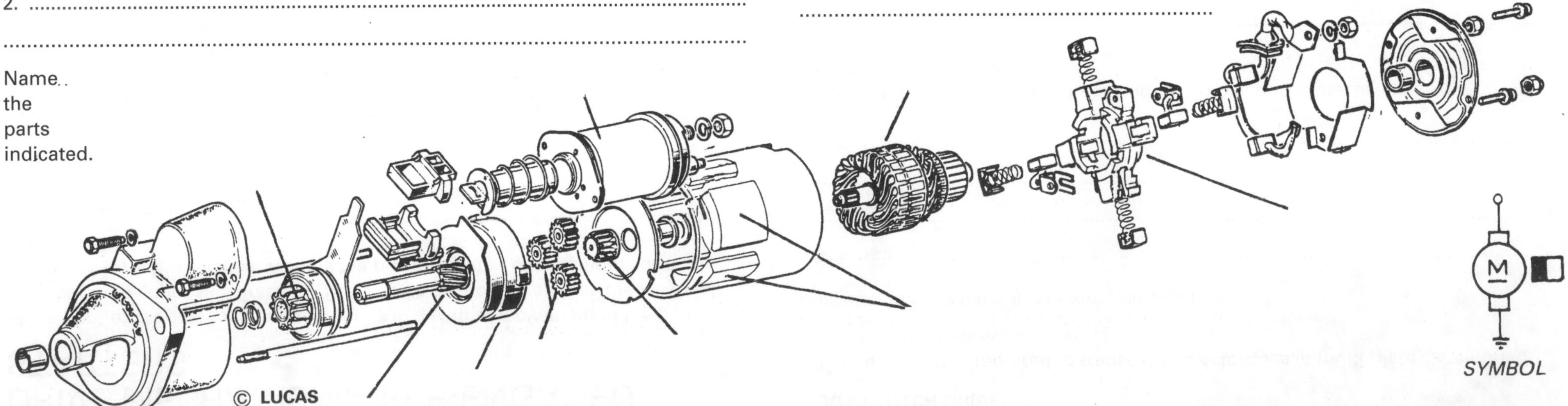

SYMBOL

AXIAL STARTERS

The starting of heavy vehicle engines, both petrol and (especially) diesel, poses special problems, mainly because of the much larger compression loads. The diesel engine also requires to be motored for longer than the petrol engine in cold temperatures.

The axial starter shown (semi-pictorially) below employs additional auxiliary winding and a holding coil as well as the main coil.

Examine a starter of this type, observe its operation and answer the questions opposite.

In the position illustrated on the wiring diagram below, the starter switch has just been closed. Show the field poles and name the parts indicated.

© **LUCAS CAV**

How does this armature design differ from others?

....

....

....

....

....

....

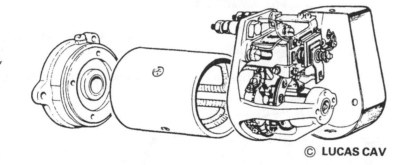

Contacts Y

Yoke

Contacts Z

Solenoid

Armature

A

1. In what position are contacts Y and Z when the starter switch is open?

....

2. Explain briefly what happens initially when the starter switch is closed.

....

....

3. Which windings are energised in this condition (shown opposite)?

....

....

4. What effect does this have on the armature?...................................

....

5. What happens to the contacts as a result of this?...........................

....

6. How does this affect the main and auxiliary windings?

....

....

7. What is the purpose of the holding coil?

....

8. What is the purpose of the spring A shown on the right-hand end of the armature?

....

....

9. Why is a multi-plate clutch used between the pinion and armature shaft?

....

....

159

COAXIAL STARTERS

The coaxial starter differs from the axial starter in that only the pinion moves axially.

How is this movement created?

....
....
....
....
....

From battery

From starter switch

At rest position

Examine a coaxial starter and answer the following questions:

1. What provision is made to cause the armature to rotate slowly at reduced voltage during initial engagement of the pinion?

....
....

2. Why is a helix required on the pinion shaft? ....

....

3. How is full voltage applied to the armature when the pinion is fully engaged?

....
....

4. What is the function of the ball locking device?

....
....
....

The drawings show the pinion engagement sequence of a CAV coaxial starter. Label the parts as required and explain alongside each drawing what is happening.

A

B

C

© LUCAS CAV

....
....
....
....
....
....
....
....
....
....
....
....
....
....
....
....
....
....
....
....
....
....
....
....
....
....
....
....
....
....

TESTING STARTER MOTORS

If the starter motor is considered suspect, for example it is cranking the engine, but only slowly, a systematic check should be carried out to determine if there is an excess voltage loss (high resistance) in the circuit.

On a four-cylinder engine the current should not exceed 140 A when cranking evenly at a speed of not less than 180 rpm.

INSPECTION

Before tests are carried out, preliminary checks that must be made are:

....

....

....

See battery testing on page 139.

In all six starter tests shown, the engine must be cranked without starting. This is achieved by:

SI engine ....

CI engine ....

Carry out a series of voltage checks to determine the condition of a starter circuit using a 0–20 V voltmeter. State the function of each check.

Show position of voltmeter for each test.

State expected and actual readings.

Test	1	2	3	4	5	6
Expected voltage						
Actual voltage						

1. Battery voltage on load

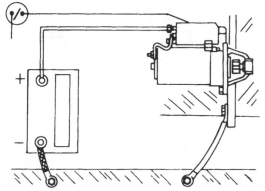

2. Voltage at solenoid operating terminal

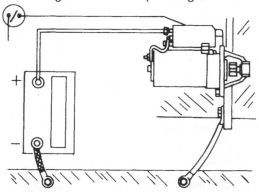

3. Voltage at starter on load

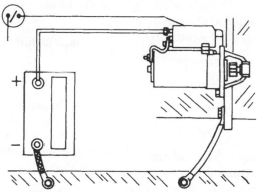

4. Voltage drop insulated link

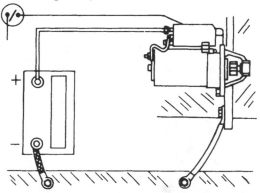

5. Voltage drop solenoid contacts

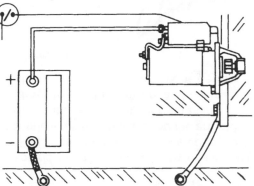

6. Voltage drop earth line

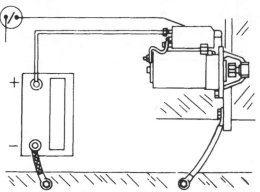

TESTING STARTER MOTORS

It is necessary to supply a very large amount of current to the starter motor to enable it to turn the engine from a stationary position. This causes a considerable voltage drop in the circuit. The starter motor is wired in such a way that it produces maximum torque immediately on turning (locked torque). This maximum torque is required to overcome the resistance to movement of the engine.

LOCK TORQUE TEST

Check the locked torque and locked current of a starter motor. Clamp starter motor in suitable test rig and attach torque arm.

Operate starter motor and obtain voltage drop, locked current and locked torque readings.

Starter make .... Model ....

Readings	Actual results	Manufacturer's specification
Locked current		
Locked torque		
Voltage drop		

Using the lock torque tester or a bench tester, other related starter tests can be carried out.

List and describe such tests.

....

....

....

....

....

....

....

TEST EQUIPMENT

For a starter light-running test, the equipment required would be:

1. ....

2. ....

3. ....

4. ....

The meter used to check the resistance of cables would be an:

....

CHECKING ARMATURES

When the starter is dismantled, an electrical tester called a 'growler' can be used to detect shorts in the armature windings. The growler is connected to the mains so that the AC supply produces magnetic lines of force around the armature similar to those produced in the starter motor when it is operating. The magnetism allows the armature windings to conduct current and the machine produces a 'humming' noise which changes tone should a fault be detected.

Describe the tests being made on the armature shown.

Show where these meters would be connected in the diagram below for a light-running test on the bench.

voltmeter   ammeter

© DELCO REMY

....

....

....

....

INSPECTION FOR FAULTS

If an external fault has been determined by testing then the starter motor must be dismantled for examination to identify the possible faults.

With the aid of workshop manuals describe the main points to be observed when:

1. removing a starter motor from the vehicle

....

....

....

....

2. dismantling a starter motor

....

....

....

....

....

....

....

What type of faults would make it necessary to consider the fitment of a replacement component?

....

....

....

Complete the table below by stating what checks should be made on the components listed to assess their serviceable condition.

COMPONENT	CHECK FOR
Armature	....
Commutator	....
Field coils	....
Brush gear	....
Bearings	....
Solenoid	....
Pinion drive	....
Clutch assembly	....
Coaxial Engagement mechanism	....
Coaxial Trigger operation	....

DIAGNOSTICS: STARTER MOTOR SYSTEM – SYMPTOMS, FAULTS AND CAUSES

State a likely cause for each symptom/system fault listed below. Each cause will suggest any corrective action required.

SYMPTOM/SYSTEM FAULTS	LIKELY CAUSE
Starter cranks engine very slowly	
There is an abnormal clonking noise when starter is operated	
There is a grating sound as pinion engages and disengages flywheel	
When starter key is turned nothing happens, but clicking sound is heard	
Starter cranks engine slowly and its torque output is low	
Starter cranks engine slowly and its current consumption is low	
Starter cranks engine slowly and its current consumption is high	
There is a very heavy clonking noise when engine is started	

SYSTEM PROTECTION DURING USE

Describe how the following should be protected from hazards during normal use or repair:

Cables ....

....

Terminals (starter-battery) ....

....

....

The starter system may have a number of inhibitor switches fitted in the circuit. What is their function?

....

....

The most common inhibitor switch is fitted to vehicles using automatic transmissions. Where is this switch fitted and what is its function?

....

....

Complete the diagram to show positions where the inhibitor switches may be placed.

What other items may include an inhibitor switch in their circuit?

....

....

....

....

....

Battery

Ignition switch

Solenoid

Starter motor

State reasons why some starter systems using a pre-engaged type drive use an additional relay as shown on the diagram:

....

....

....

Name the main parts on the diagram opposite and discuss its operation.

GOOD WORKING PRACTICE

List the general rules/precautions to be observed to ensure efficient testing, overhauling and repairing of starter motor systems when:

1. Removing and refitting ....

....

....

....

....

....

2. Using test equipment ....

....

....

3. Bench testing correctly..

....

....

Chapter 9

Engine – Mechanical Details, Inspection and Testing

Engine components – their basic forms and locations	167	Connecting rod – small end	191
Functional requirements of engine components	168	Connecting rods – big end bearings	192
Cylinder heads	169	Connecting-rod alignment and inspection	193
Nut-tightening sequences	170	Four-cylinder in-line crankshafts	194
Camshafts, poppet valves	171	Cylinder numbering and firing order	195
Push-rods and rockers (valve operation)	172	Four-cylinder vee crankshafts	195
Valve clearance	173	Six-cylinder in-line crankshafts	196
Valve guides, valve springs	174	Built-up crankshafts, six-cylinder vee crankshafts	196
Valve spring retention	174	Vee-eight crankshafts	197
Overhead camshaft valve arrangements	175	Attachment of components to the crankshaft	198
Using shims to adjust valve clearance	176	Crankshaft alignment – principles	199
Camshaft drive arrangements, gear drives	177	End-float – crankshaft, inspection	200
Chain drives	177	End-float – camshaft	200
Automatic mechanical tensioner	177	Engine balance	201
Overhead camshaft belt drives	178	Crankshaft torsional vibration	202
Variable valve timing	179	Crankshaft vibration dampers	202
Self-adjusting tappets, cylinder head faults	180	Crankshaft vibration frequency	203
Cylinder blocks	181	Engine mountings	203
Sealed-for-life cylinder block and head	182	Overhauling/reconditioning engines	204
Crankcases	182	Main statutory requirements	204
Cylinder liners	183	Diagnostics	205
Fitting cylinder liners	185	Engine diagnosis – lack of performance	207
Checking cylinder bore wear	186	Testing performance using chassis dynamometer	208
Piston construction, inspection	187	Exhaust gas analyser	209
Piston rings	188	Engine performance testing	210
Inspections – piston rings	189	Performance-modifying parts	211
Gudgeon pins	190	Safe working practice	212

ENGINE COMPONENTS – THEIR BASIC FORMS AND LOCATIONS

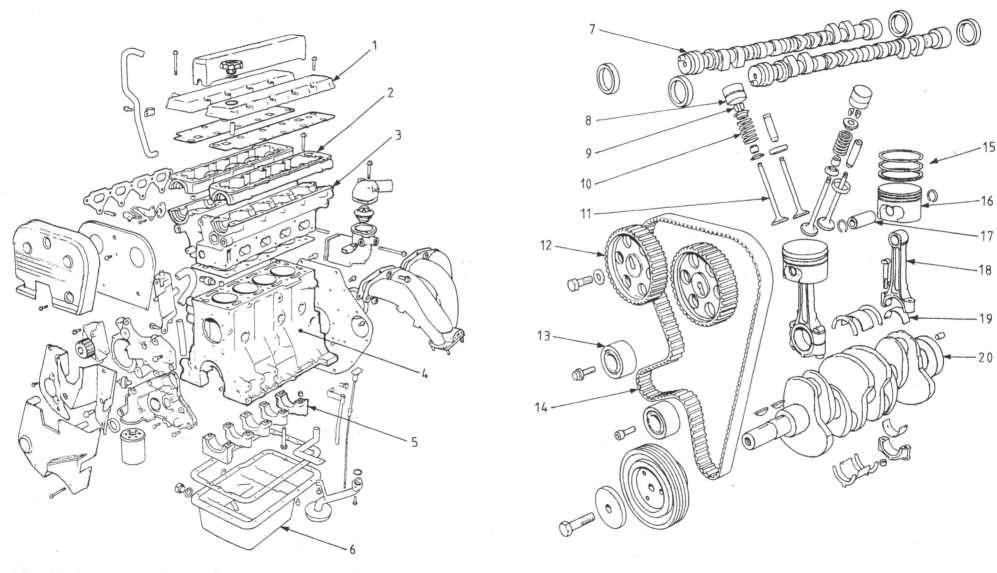

PURPOSE AND FUNCTIONAL REQUIREMENTS OF ENGINE COMPONENTS

Name the parts numbered on the previous page and briefly state their basic function.

Part	Function
1.	....
2.	....
3.	....
4.	....
5.	....
6.	....
7.	....
8.	....
9.	....
10.	....

Part	Function
11.	....
12.	....
13.	....
14.	....
15.	....
16.	....
17.	....
18.	....
19.	....
20.	....

CYLINDER HEADS

The cylinder head is mounted on top of the cylinders. It confines the pressure of combustion and directs it down on to the piston. The head also provides water-cooling passages, inlet and exhaust passages and supports the valve gear.

Identify the important features of the cylinder heads shown.

UNIFLOW HEAD

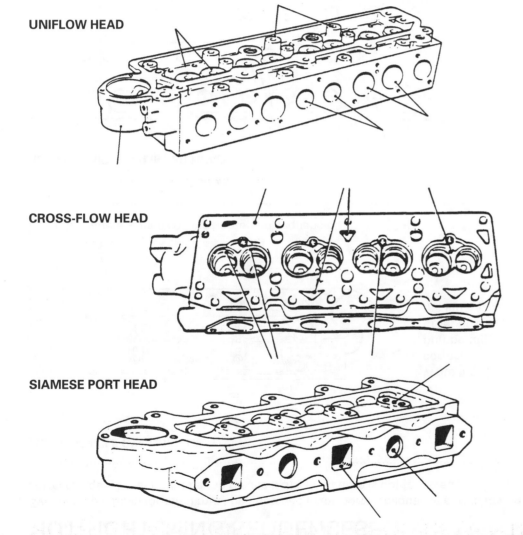

CROSS-FLOW HEAD

SIAMESE PORT HEAD

Identify the important features of the cylinder heads shown with reference to:

(a) number of valves, (b) port arrangements, (c) spark plug location.

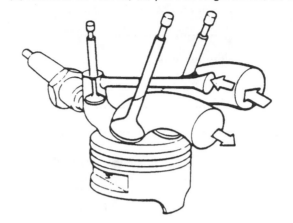

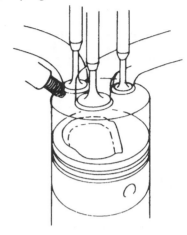

(a) .... (a) ....

(b) .... (b) ....

(c) .... (c) ....

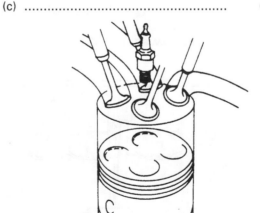

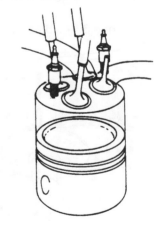

(a) .... (a) ....

(b) .... (b) ....

(c) .... (c) ....

NUT-TIGHTENING SEQUENCES

Cylinder head bolts (or nuts) must be tightened in the correct sequence and to the correct torque. What might be the results if this procedure is not followed?

....

LIGHT-VEHICLE CYLINDER HEAD

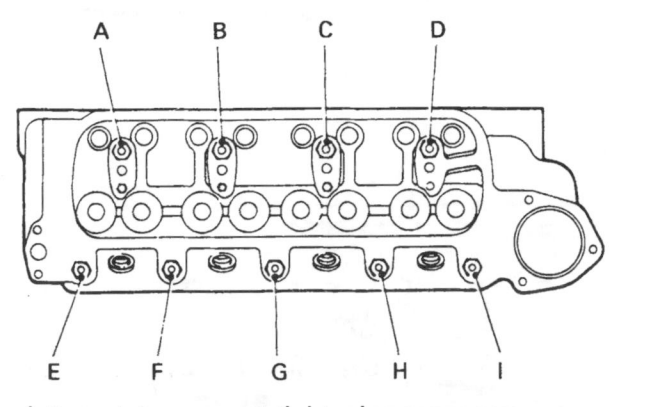

Write the correct numbering order next to the letters.

Using the letters state a correct tightening sequence:

....

Typical torque setting would be ....

HEAVY-VEHICLE CYLINDER HEAD

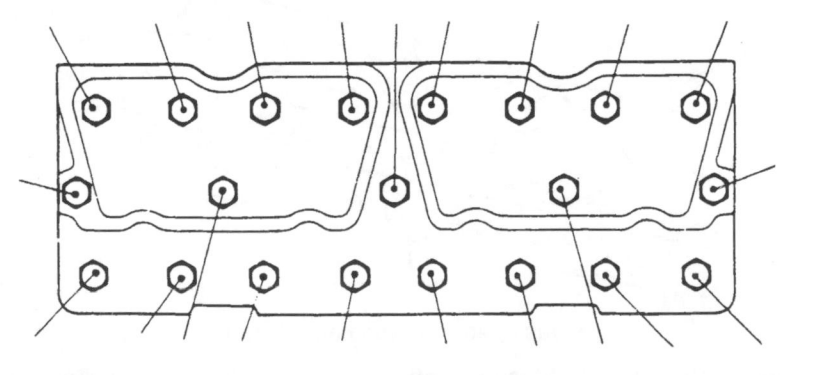

Number the bolts in the correct tightening sequence.

Typical torque setting would be ....

On older type engines when tightening the cylinder head bolts it is usually only necessary to tighten, in the correct sequence, for example all bolts to 50 Nm (first stage) and then all bolts to 75 Nm (second stage).

A more common modern method is to tighten, first by using a torque wrench and then by using an angle measuring tool.

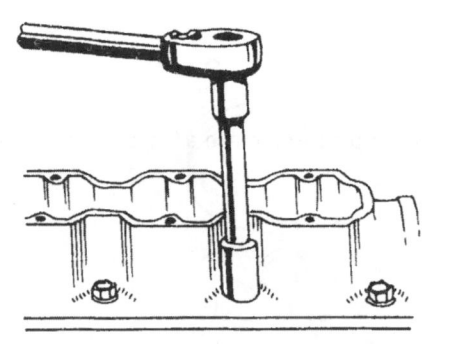

First stage: Torque wrench   Second stage: Angle tool

Obtain from 'Technical Data' the cylinder head tightening stages of torques and angles turned.

Vehicle 1) _Ford Escort 1.6_   Vehicle 2) ....

Stage 1 _25 Nm_   Stage 1 ....

2 _55 Nm_   2 ....

3 _90 degrees_   3 ....

4 _90 degrees_   4 ....

Vehicle 3) ....   Vehicle 4) ....

Stage 1 ....   Stage 1 ....

2 ....   2 ....

3 ....   3 ....

4 ....   4 ....

5 ....   5 ....

6 ....   6 ....

CAMSHAFTS

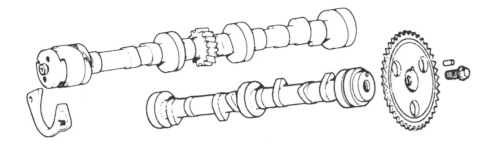

The function of the camshaft is to operate the valves, and frequently it forms a convenient mounting point for various auxiliary drives. Its position on the engine and the actual shape of the cams themselves vary according to individual manufacturers' requirements.

Name three auxiliaries that could be driven by the camshaft:

....

Suggest one important advantage of a side-mounted camshaft as compared with an OHC layout:

....

....

What features dictate cam shape?

....

....

State the method used in the above sketch to locate the drive gear on the shaft:

....

Suggest an alternative method of accurate location of the cam drive gear on the shaft:

....

POPPET VALVES

In almost all four-stroke engines, the poppet valve controls the flow of gas into and out of the cylinder.

On modern vehicles the valve seat angle usually is ....

On some CI engines the inlet valve seat angle may be ....

The diagram shows the correct contact of valve to seat.

Shade on the valve the area of valve–seat contact.

State the effects of

(i) excessive clearance ....

....

....

(ii) insufficient clearance ....

....

Show, with a sketch similar to the one above, a sectional view of a valve seat insert fitted into the cylinder head.

State the required amount of interference fit ....

From what alloys are valve seat inserts made?

....

....

Why do engines using lead-free fuel need to have hard-alloy valve seat inserts?

....

PUSH-RODS AND ROCKERS (VALVE OPERATION)

On overhead valve engines with the camshaft positioned in the cylinder block, the valves are operated via cam followers, tappets, push-rods and rockers.

Indicate on the drawing the method of valve adjustment and the means of lubricating the valve mechanism.

On some modern push-rod-operated engines, the rocker shaft has been eliminated.

Such a mechanism is shown below.

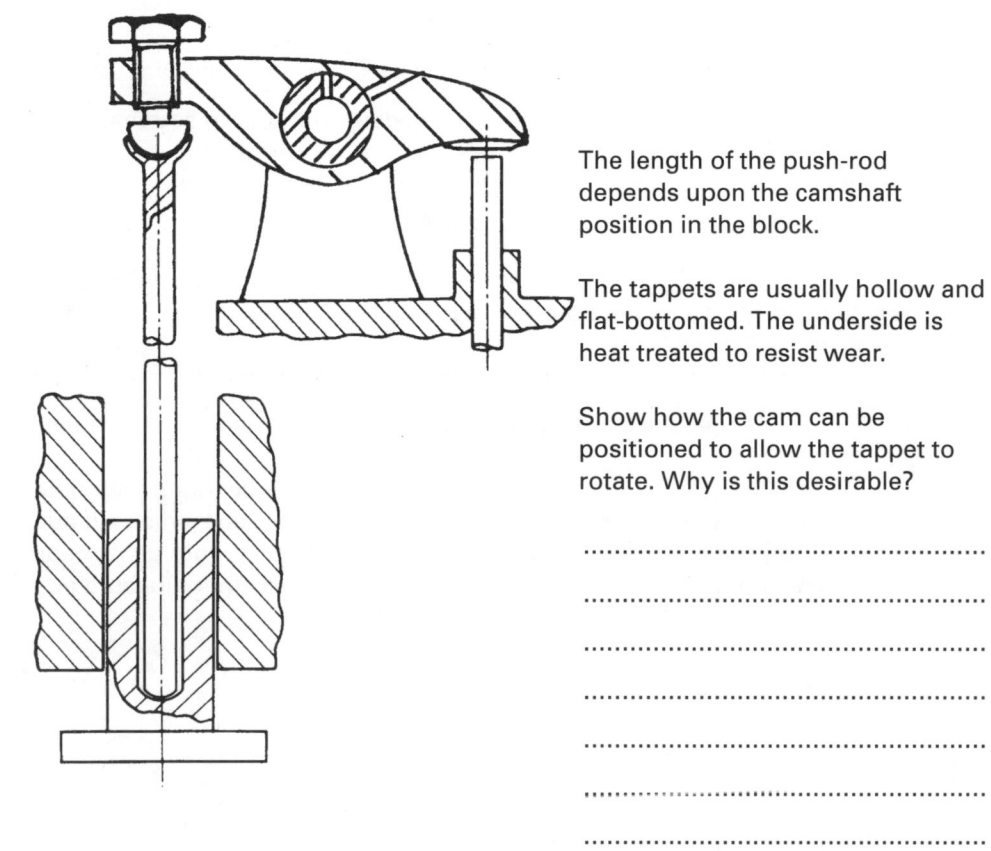

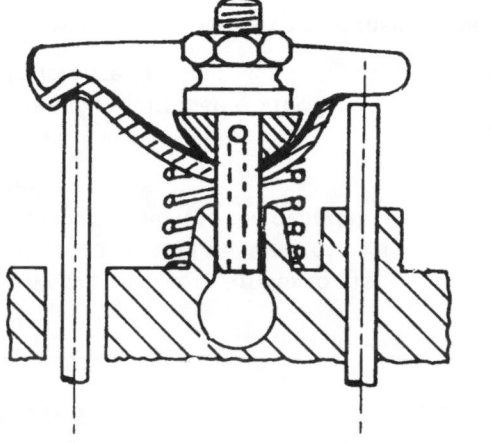

The length of the push-rod depends upon the camshaft position in the block.

The tappets are usually hollow and flat-bottomed. The underside is heat treated to resist wear.

Show how the cam can be positioned to allow the tappet to rotate. Why is this desirable?

....
....
....
....
....
....
....

Indicate the method of adjustment and show how oil is supplied to the rocker.

What is the advantage of using this type of mechanism?

....
....
....

With this type of design, the rocker arm tends to contact the valve at an angle and at speeds over about 2000 rev/min will cause the valve to slowly rotate.

Why is this rotation an advantage?

....
....
....

VALVE CLEARANCE

The valves, when operated, require clearance. There are two reasons for this:

(a) ....

(b) ....

INVESTIGATION

Examine an in-line four-cylinder OHV engine with the rocker cover removed.

Make of engine .... Model ....

Using I for inlet and E for exhaust show the valve order along the head	1	2	3	4	5	6	7	8
State an alternative valve order								

State the engine's valve clearances

Exhaust valve .... Inlet valve ....

Describe a typical procedure for adjusting the valve clearance on the engine examined.

....

....

....

....

....

....

....

What are the main advantages of hydraulic tappets?

....

....

METHODS OF VALVE ADJUSTMENT

Name the type of adjustment shown below and state precisely where clearance is measured. Indicate the position with an arrow.

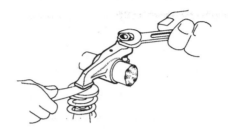

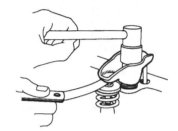

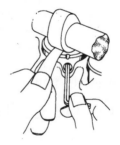

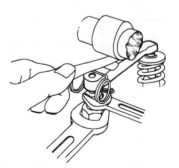

VALVE GUIDES

The valve stem is located in a guide which may be integral with the head or in the form of a sleeve which is an interference fit in the head or block.

Name the types of valve guides shown and state the most probable material used in each case.

(a) 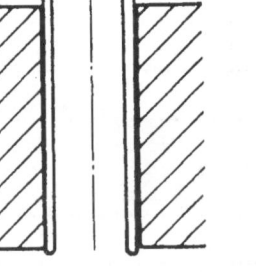 **(b)** 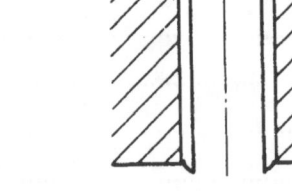 **(c)**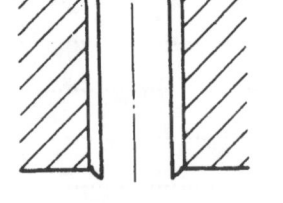

Type ....   Type ....   Type ....

Material ....   Material ....   Material ....

How would excessive clearance be brought back to standard in the following types?

(a) ....

....

(b and c) ....

....

....

State a typical valve stem clearance ....

What would be the effect of:

Insufficient stem clearance? ....

....

Excessive stem clearance? ....

....

VALVE SPRINGS

The valve is opened, either directly or indirectly, by a cam on the camshaft and is usually closed by a coil spring.

The valve spring may be a single spring having either:

(a) uniform coils,
(b) the coils wound closer together at the cylinder head end,
(c) two springs, one inside the other.

What is meant by valve bounce? ....

....

Give the advantages of using type:

(b)...

....

....

....

(c)...

....

....

....

VALVE SPRING RETENTION

The most common arrangement is by split collets.

Remove a valve from a cylinder head. examine it and complete the sketch opposite. Show clearly the method of spring attachment and stem oil-sealing arrangement. What does the oil seal prevent?

....

....

....

....

....

....

Two alternative types of valve spring retention are:

....

OVERHEAD CAMSHAFT VALVE ARRANGEMENTS

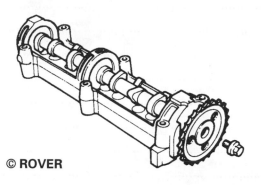

© **ROVER**

With the overhead camshaft design, the most common arrangement is to allow the cam to operate directly on the tappet block which is in direct contact with the valve.

Draw the adjusting shim in its correct position.

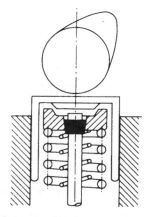

The above design shows a method of obtaining the valve clearance by using shims of varying thickness. *Note*: When the engine has been assembled the clearance cannot be adjusted

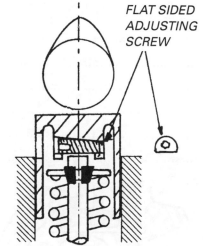

FLAT SIDED ADJUSTING SCREW

This diagram shows a design where it is not necessary to remove the camshaft when adjusting valve clearance.

Indicate how adjustment is made.

An alternative design would be to have the shim in a recess on top of the tappet.

An alternative arrangement is to have the cam operate the valve via finger levers.

Give an advantage and a disadvantage of this arrangement compared with the other two arrangements on this page:

....

....

....

Indicate the provision for valve adjustment and name the parts.

© **FORD**

The oil seal is an alternative design to the one on the previous page.

What are the main advantages of using an OHC design compared with side camshaft?

....

....

....

....

....

ADJUSTING VALVE CLEARANCE ON OVERHEAD CAMSHAFT ENGINES USING SHIMS

Indicate on the drawing where the clearance is measured and what is adjusted.

Describe a typical valve checking procedure.

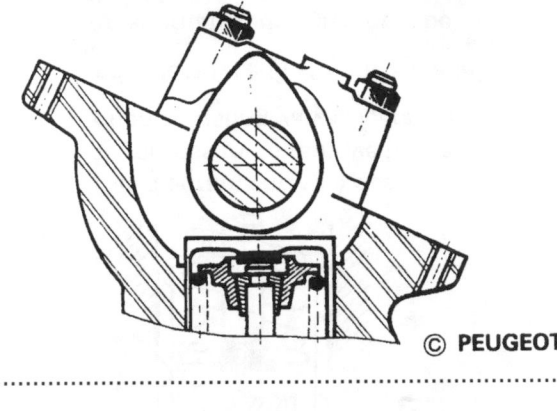

© PEUGEOT

....
....
....
....
....
....
....
....
....
....
....
....
....
....
....
....
....
....

METHOD OF CALCULATING REQUIRED SHIM THICKNESS

Typical example:

Measured clearance	0.21 mm
Manufacturer's clearance	0.30 mm
Difference (smaller)	–0.09 mm
Existing shim thickness	= 2.81 mm

New shim thickness to be

$$2.81 - 0.09 = 2.72 \text{ mm}$$

(or fit nearest available thinner size)

Note on OHC engines: clearance decreases as engine wears.

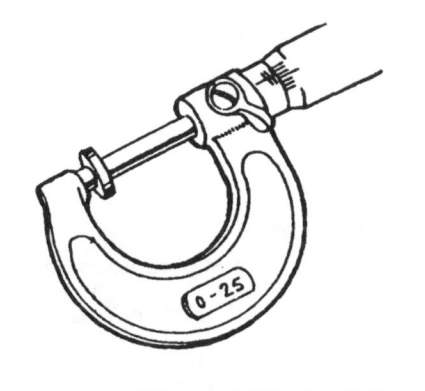

METHOD OF MEASURING TAPPET SHIM

Carry out this procedure on an engine and complete the table below:

Type of engine ....

Specified valve clearance. Inlet .... Exhaust ....

Valve No.	1	2	3	4	5	6	7	8
Measured clearance								
Specified clearance								
Difference								
Measured shim thickness								
Calculation for new shim								
New shim thickness								

CAMSHAFT DRIVE ARRANGEMENTS

Camshafts may be driven by gears, chains or belts.

Engines having camshafts positioned in the crankcase may be driven by:

....

The rotational speed of the crankshaft is .... that of the camshaft.

The speed ratio of crankshaft to camshaft is ....

How does this ratio affect the size of the gears?

....

....

When the crankshaft rotates at 50 rev/s its circumferential (linear) speed is 10 m/s.

What would be the circumferential speed of the camshaft? ....

GEAR DRIVES

Show in correct size relationship a camshaft gear driven from the crankshaft shown.

In certain cases the camshaft gear is made of a fibre material.

The reason for this is ....

....

Camshaft rotation when gear driven is...

....

When chain driven the camshaft rotation relative to the crankshaft is:

....

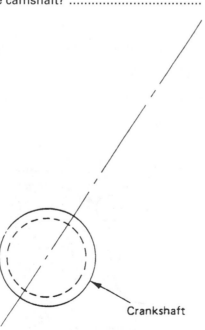

Crankshaft

CHAIN DRIVES

Add the chains to each sketch.
Name the arrowed parts and show direction of rotation.

Camshaft

....

....

....

....

TYPE (a) ....   (b) ....

AUTOMATIC MECHANICAL TENSIONER

Snail cam tensioner

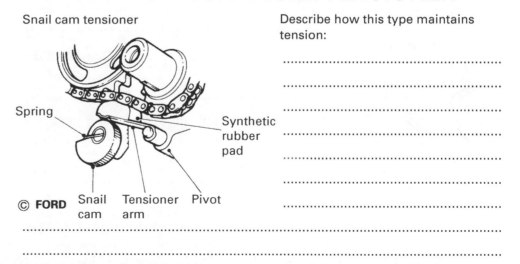

Spring

Synthetic rubber pad

© **FORD**   Snail cam   Tensioner arm   Pivot

Describe how this type maintains tension:

....

....

....

....

....

....

....

....

OVERHEAD CAMSHAFT BELT DRIVES

Add the belts to each sketch.

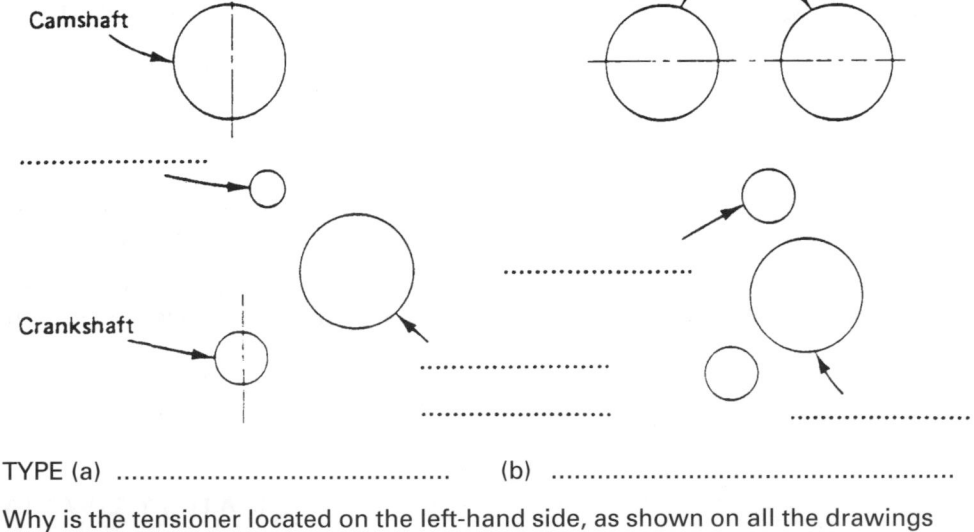

TYPE (a) .... (b) ....

Why is the tensioner located on the left-hand side, as shown on all the drawings on this page?

....

....

Why should the crankshaft not be rotated when the belt is removed?

....

....

What damage can occur if the timing belt fails when the engine is running?

....

....

State the recommended belt change mileage for three vehicles:

1. ....

2. ....

3. ....

Describe a Timing Belt removal and refitting procedure.

TYPE OF VEHICLE ....

SPECIAL PRECAUTIONS

1. ....

2. ....

....

3. ....

4. ....

5. ....

6. ....

REMOVAL PROCEDURE

1. ....

....

2. ....

....

3. ....

4. ....

5. ....

6. ....

....

7. ....

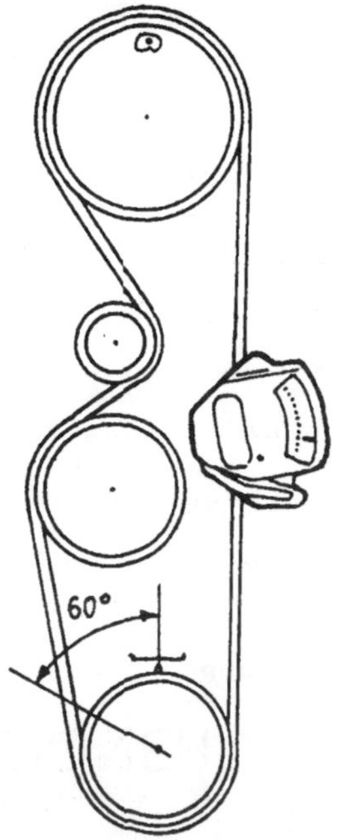

FITTING NEW BELT

1. ....

2. ....

3. ....

4. ....

5. ....

6. ....

7. ....

8. ....

Sketch shows position of timing belt tension gauge.

Note: When checking tension, crankshaft must be rotated 60° anticlockwise from tdc (Ford Escort).

VARIABLE VALVE TIMING

One way of achieving both the performance of a high output engine and the low torque output of a standard engine is to use Variable Valve Timing. There are TWO basic designs, the simplest is used on twin camshaft engines and only advances the inlet camshaft as the engine speed increases. The second type allows all the valves to operate on conventional cam profiles and then to change to high performance high lift cam profiles.

VARIABLE INLET VALVE TIMING

At low speed the inlet camshaft acts as a normal camshaft having a small valve overlap, then as the engine speed increases the camshaft can be advanced up to 12.5° (25° of crankshaft rotation). The valve open period remains the same.

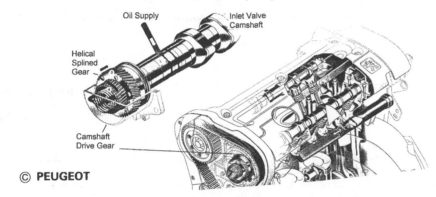

© **PEUGEOT**

Describe the action of the gears shown to advance the inlet valve timing.

....
....
....
....
....
....
....

HONDA VARIABLE VALVE TIMING AND VALVE LIFT

The Honda system allows the valves to operate on two sets of cams and so provide the valve timing of a standard engine at low speed and of a high output engine at high speed.

The valve mechanism for each pair of valves contains a third (mid) rocker arm which does not engage at low speed. Hydraulic pistons (spool valves) are built into the three rocker arms and when increasing oil pressure moves the spool valve the three rocker arms lock up and work as a single rocker operated by the central HIGH LIFT cam lobe. This central rocker is spring tensioned to prevent lost motion, when it is not locked at low speed.

Describe which cams are driving the rocker arms in the drawings below.

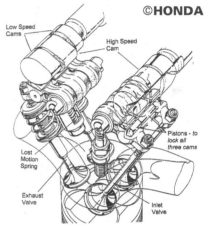

©**HONDA**

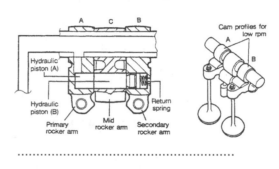

....

....

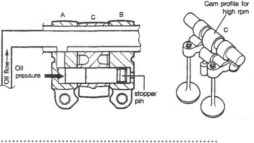

Complete the block diagram to show what factors govern the operation of the solenoid valve.

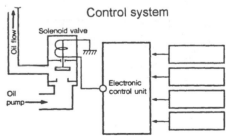

....

What are the valve lift values of the two cam profiles? ....

179

SELF-ADJUSTING TAPPETS

Describe the operation of the hydraulic tappet shown:

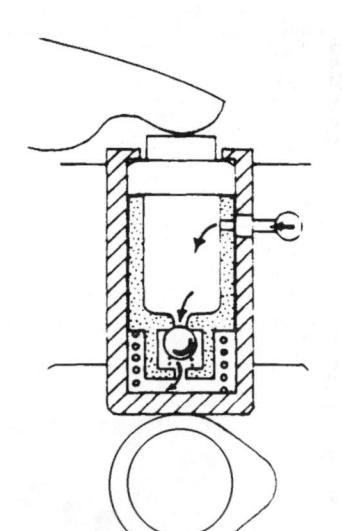

....

....

....

....

....

....

....

List common faults found during service associated with:

(a) camshafts and tappets

1. ....

2. ....

3. ....

4. ....

5. ....

6. ....

7. ....

8. ....

(b) timing drives

1. ....

2. ....

3. ....

CYLINDER HEAD FAULTS

List eight common service faults associated with cylinder heads:

1. ....

2. ....

3. ....

4. ....

5. ....

6. ....

7. ....

8. ....

....

How can the distortion of the head-to-block machined face be rectified?

....

....

....

....

....

List six common service faults associated with the valves in the cylinder head:

1. ....

2. ....

3. ....

4. ....

5. ....

6. ....

CYLINDER BLOCKS

On modern engines the 'cylinder block' is the term usually given to the combination of the block itself which carries the cylinders, and the crankcase, which supports the crankshaft, that is, the basic 'framework' of the engine.

A separate cylinder block is commonly used for motor-cycle, air-cooled and heavy commercial vehicle engines.

Give a more correct definition of the cylinder block:

....

....

....

....

....

State in each case TWO reasons for adopting this separate cylinder block type of construction:

Single-cylinder motor-cycle engines

1. ....

   ....

2. ....

   ....

Air-cooled multi-cylinder engines

1. ....

   ....

2. ....

   ....

Heavy commercial vehicles

1. ....

   ....

2. ....

   ....

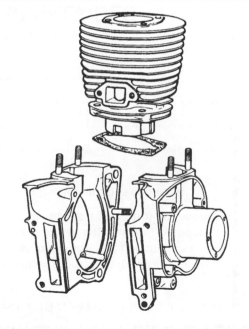

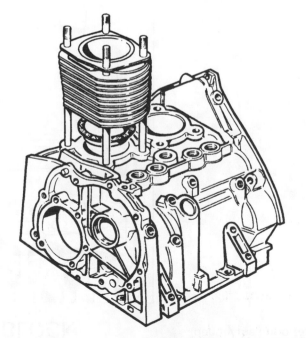

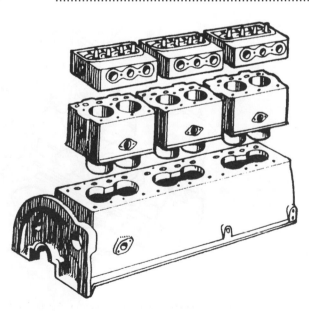

SEALED-FOR-LIFE CYLINDER BLOCK AND HEAD

On some compression-ignition engines (an example is shown opposite) and small two-stroke engines, the cylinder block and head form a complete casting.

What are the advantages of such a design?

....

....

....

....

....

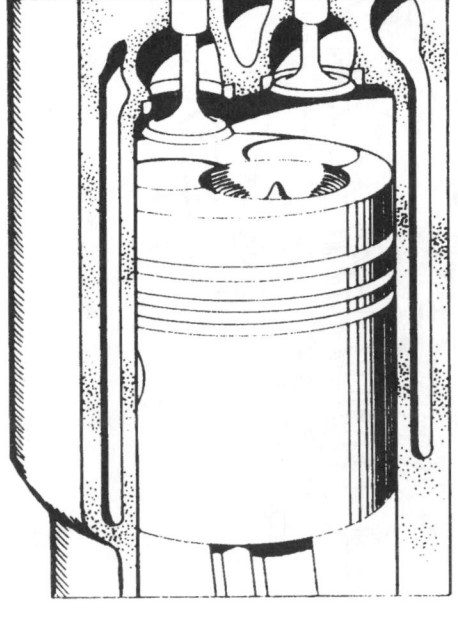

CRANKCASES

The engine crankcase may be a separate unit supporting the crankshaft and camshaft or be an integral part of the cylinder block.

Describe the constructional details of:

(a) Separate crankcases.

....

....

(b) Combined cylinder block and crankcase, where the crankcase lower face is in line with the centre line of the main bearings, or is extended below the centre line of the main bearings as shown opposite.

....

....

....

Identify the important parts of the integral crankcase shown below.

© **VAUXHALL**

How are the effects of distortion and vibration created in the crankcase minimised?

....

....

....

....

....

Why are there always core (welch) plugs fitted to the side of the cylinder block of a multi-cylinder water-cooled engine?

....

....

....

CYLINDER LINERS

Cylinder liners are fitted into some engine blocks as shown opposite.

What purpose do they serve when compared with integral type cylinders?

....

....

....

State two types of cylinder liner material:

1. ....

2. ....

List the main advantages of fitting cylinder liners:

....

....

....

....

What are the main causes of rapid cylinder bore wear?

....

....

....

....

What reasons may cause substantial variations in wear between different cylinders on the same engine?

....

....

....

....

Identify the type of liners shown.

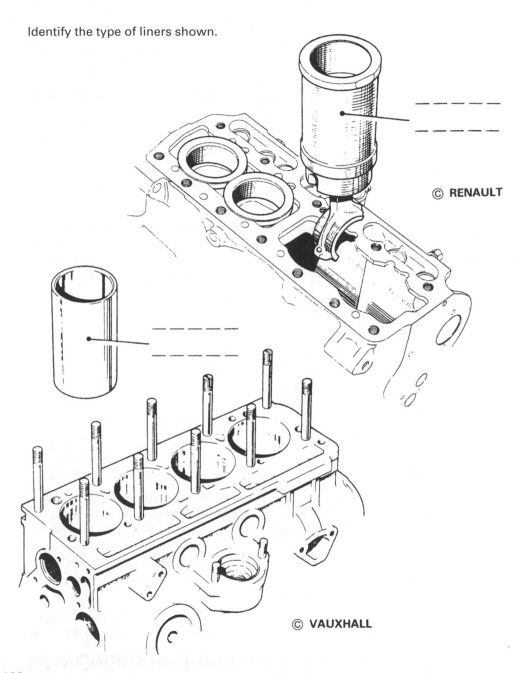

© RENAULT

© VAUXHALL

Dry Cylinder Liners

These are pressed directly into the cylinder bore.

Two types are used.

Describe their fitting procedures.

1. Interference fit

....

....

....

....

....

....

....

Complete the diagrams below to show dry cylinder liners.

Interference fit liner

2. Slip fit

....

....

....

....

....

....

....

....

Slip fit liner

Wet Cylinder Liners

This type of liner has its outer surface in direct contact with the cooling water. There are two types in common use:

Name the items indicated on the diagrams below.

(a) The liner's top shoulder is located in a recess in the cylinder block faces while the lower end is a push fit into the lower part of the block. What are the advantages of dry cylinder liners?

....

....

....

....

....

....

Type commonly fitted in CI engines

(b) This type block has an 'open deck' layout with the liners inserted into spigots rising from the lower part of the block. These liners may be 'cast-in' on some engines.

On what type of engine is this design most common?

....

....

....

....

Type commonly fitted in OHC engines

FITTING CYLINDER LINERS

1. INTERFERENCE FIT LINERS

Examine an interference fit cylinder liner and cylinder block into which it is to fit. If possible measure the cylinder bore diameter and external cylinder liner diameter and determine the interference.

Cylinder bore diam.	External liner diam.	Interference

What checks should be made on the cylinder block before fitting the liner?

....

....

What should be done to assist the liner to be fitted into the bore?

....

....

How must the liner be finished off when fitting is complete with regard to:

(a) the cylinder block face?

....

....

(b) its internal diameter?

....

....

2. WET LINERS

Describe what is being shown in the sketch.

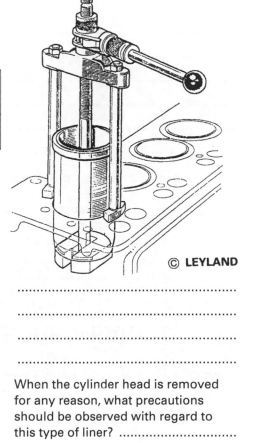

© **LEYLAND**

....

....

....

....

When the cylinder head is removed for any reason, what precautions should be observed with regard to this type of liner? ....

....

....

The sketch shows a method of holding down the cylinder liner during servicing.

If such an extension piece was not available, what alternative method could be used?

....

....

....

....

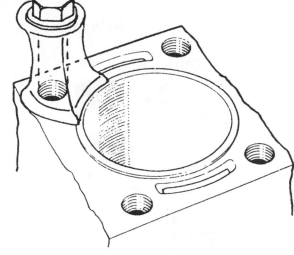

Name the main parts

Describe a typical wet cylinder liner fitting procedure:

....

....

....

....

....

....

....

....

....

State the amount of protrusion when fitted to provide 'nip':

Actual protrusion ....   Manufacturer's specification ....

List faults associated with cylinder liners commonly found during service:

1. ....
2. ....
3. ....
4. ....
5. ....
6. ....

CHECKING CYLINDER BORE WEAR

The diameter of cylinder bores may be checked by:

....

An internal micrometer takes direct readings but requires a sensitive touch to obtain accurate readings. When a cylinder bore gauge is used it must be calibrated by using either a ring gauge or an external micrometer.

Describe how the cylinder bore wear may be checked, using such a gauge:

....
....
....
....
....
....

Having determined the original bore diameter and maximum wear reading, state two other values that can be measured.

....
....
....

List common faults found during service associated with cylinder blocks:

1. ....
2. ....
3. ....
4. ....
5. ....
6. ....

The cylinder bore gauge converts the horizontal movement of the spring-loaded plunger into a vertical movement, which is transferred by a push-rod (in the gauge shaft handle) to a dial test indicator clamped to the top of the handle.

Name the major parts.

Setting cylinder bore gauge

to read ....

Checking cylinder

PISTON CONSTRUCTION

The piston forms a sliding gas-tight (and almost oil-tight) seal in the cylinder bore and transmits the force of the gas pressure to the small end of the connecting rod. In achieving this the piston must form a bearing support for the gudgeon pin and take side thrust loads created by the angular displacement of the connecting rod.

Name the various parts of the piston below and indicate using arrows the main side thrust loads created by the connecting rod.

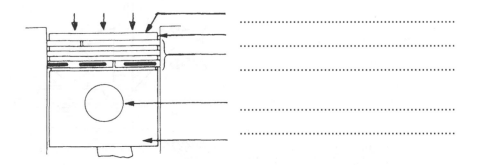

....

....

....

....

....

The shape, size, strength and material of a piston are largely dependent on the type of engine in which it is to be used, for example, spark-ignition, compression-ignition, type of combustion chamber etc.

List some of the major points that must be considered when designing pistons.

....

....

....

....

....

....

....

....

It is necessary to design pistons with operating clearances that vary according to their position on the piston. This is because of the temperature variation within the piston when it is operating.

The diagram shows how the temperature gradients vary on the crown and sides of a typical CI engine piston. Describe how differing temperatures affect expansion.

....

....

....

....

....

....

....

....

....

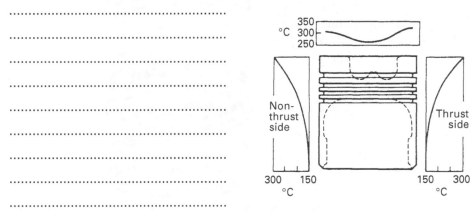

INSPECTION

To determine variation in piston diameter.

Measure, using a micrometer, two pistons of a similar size but different materials at the positions shown below.

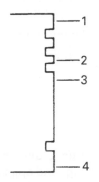

	Material	
Micro-meter position	....	....
	....	....
1.		
2.		
3.		
4.		

PISTON RINGS

What are the three basic functions of a piston ring?

1. ....

2. ....

....

....

3. ....

....

Petrol engines usually have pistons with two compression rings and one oil-control ring above the gudgeon pin.

Identify the rings shown in the sketch below.

....

....

....

....

....

COMPRESSION RINGS

The top ring is usually of a plain rectangular section, with often the outer edge either chromium or molybdenum plated.

Sketch a 'second' compression ring cross-sectional shape which is a different design from the one shown above.

OIL-CONTROL RINGS

This type of ring should glide over the oil film as the piston moves up the cylinder yet scrape off all but a thin film of oil when descending.

The oil scraper ring above the gudgeon pin on older types of engines is usually of a slotted design. Describe the ring's features.

Draw a sectioned view of a slotted scraper ring in the piston section below.

....

....

....

....

....

....

OIL SCRAPER RINGS

These are designed to exert a greater radial force on the cylinder wall than the above types. Describe the fabricated construction shown.

Compression rings

Fabricated oil control ring

....

....

....

....

The sketch below shows a section of a steel rail multi-piece ring.

HEPOLITE S E RING

INSPECTIONS

PISTON RINGS

Measure the individual gaps of a set of rings in their cylinder bore.

Assemble the rings on the piston and measure the ring groove clearance.

Record your findings in the table below.

Engine make .... Model .... Cylinder diameter ....

Ring type and position on piston	Ring gaps		Groove clearance		Serviceability
	Actual	Recommended	Actual	Recommended	
....					....
....					....
....					....
....					....

The sketches show where readings should be taken.

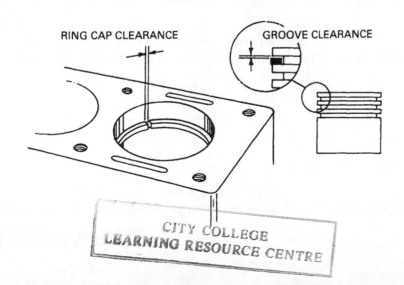

RING CAP CLEARANCE   GROOVE CLEARANCE

The sketch below shows recommended positions of the ring gaps when fitted on to the piston.

In relation to the ring gaps what must you NOT DO when fitting piston rings?

....

....

....

....

....

Thrust side

Top compression

Oil control

2nd compression

Using a special tool similar to the one shown, fit the piston into the cylinder bore.

State the procedure that should be observed:

....

....

....

....

State the approximate radial pressure of the following types of rings:

Compression ....   Normal oil scraper ....

Special oil scraper ....   Internally stepped ....

After a period in service, piston ring grooves can become blocked with deposits. What are these deposits?

....

....

GUDGEON PINS

The gudgeon pin connects the piston to the connecting rod and has to withstand the shock loads created by the forces of combustion.

....

....

....

....

The pin must also be prevented from moving sideways and scoring the cylinder walls. There are four basic ways in which this can be done.

1. Gudgeon pin fully floating held with circlips fitting into the gudgeon-pin bosses.

....

....

....

....

Describe how the gudgeon pin should be fitted to the piston and connecting rod:

....

....

....

....

....

....

....

Complete the sectioned diagram by showing the circlips in position.

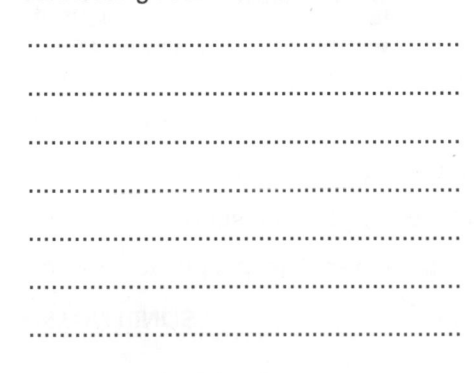

2. A fully floating gudgeon pin may be fitted with end-pads.

The diagram, right, shows a sectioned view of a gudgeon pin fitted with end-pads.

State from what material the end-pad may be made. ....

What is the function of the end pads?

....

....

....

In both these types the gudgeon pin is free-floating in the connecting-rod small end. The small end is usually fitted with a bush.

On what type of engines would the gudgeon-pin ends be sealed by a plain disc?

....

What is the reason for this sealing?

....

....

3. Gudgeon pin held in small end with a clamp-bolt (pin is a thermal fit in piston).

Why is this type not popular on modern engines?

....

....

....

....

....

4. Gudgeon-pin interference fit in small end, free in piston. **Note:** a small end bush is not fitted)

Describe the method of fitting the gudgeon pin and piston to the connecting rod on the above type.

....

....

....

....

....

....

GUDGEON PIN OFFSET

Solid skirt pistons require greater running clearances than other types and so can produce noisy 'piston slap' when the engine starts up from cold. This effect tends to be more pronounced on the modern shorter stroke type engines.

What causes piston slap?

....

....

....

....

To reduce piston slap, the gudgeon pin may be offset about 1.5mm from the piston centre line towards the maximum thrust side of the cylinder. This allows the piston to tilt during the compression stroke so that the skirt is touching the maximum thrust side of the cylinder wall and as the piston rocks when it moves onto the power stroke the slapping effect is controlled.

Show the position of the gudgeon pin offset on the drawing below, by drawing centre lines and the gudgeon pin.

MAXIMUM THRUST SIDE

d.o.r.

Why are gudgeon pins normally hollow?

....

....

CONNECTING ROD

SMALL END

The small end of the connecting rod locates the gudgeon pin.

In types 1 and 2 (page 190) the gudgeon pin is fully floating and is fitted with a bush which requires lubrication. Type 3 is a clamp type and type 4 is an interference fit; in these types the pin does not move in the rod and so no lubrication is required.

The sketch shows the little end of the connecting rod.

Describe how it is lubricated.

....

....

....

....

CONNECTING RODS – BIG END BEARINGS

Conventional type

Oblique type

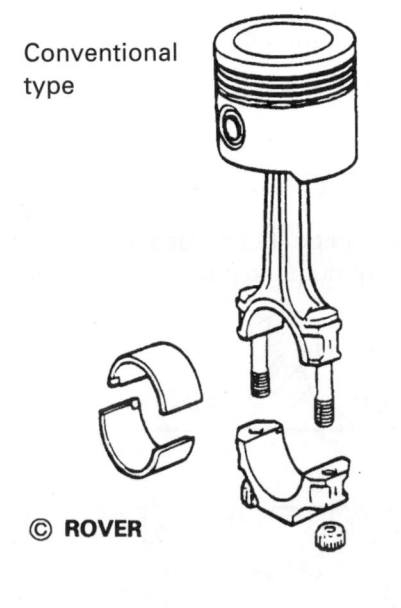
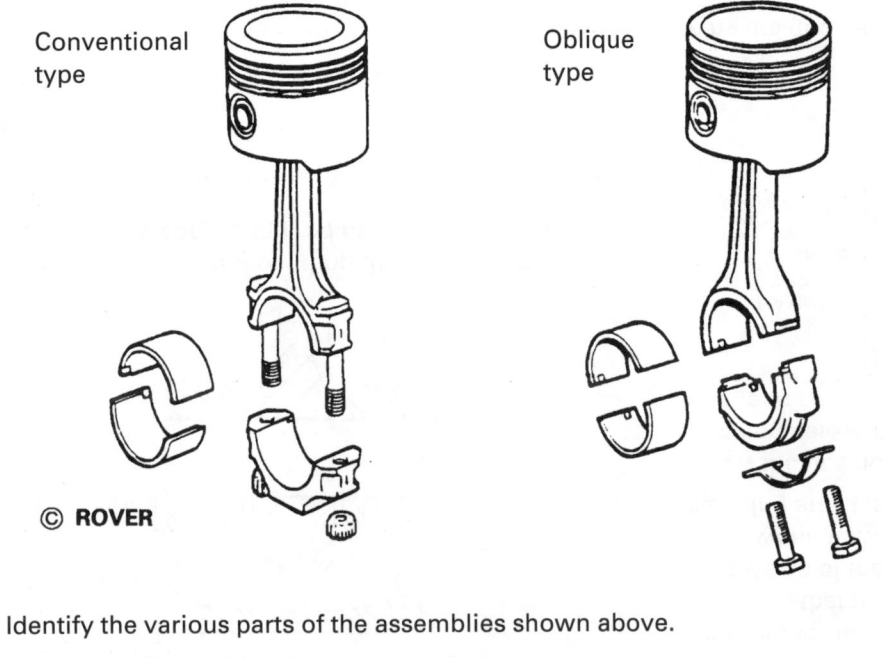

© **ROVER**

Identify the various parts of the assemblies shown above.

Describe how the bearings are:

(a) located ....

....

....

(b) prevented from rotating ....

....

....

....

What other type of bearings may be used for big ends, particularly for motor cycles? ....

The big end locates on the crankshaft and via the connecting rod enables the reciprocating motion of the piston to be transferred into the rotary motion of the crankshaft.

What is the reason for designing the big end so that the cap face split line is at an oblique angle?

....

....

....

What is the disadvantage of splitting the big end at an angle?

....

....

....

Explain how oil is supplied to the big end bearings and how they supply oil to the cylinder walls:

....

....

....

....

....

The rod must withstand the compressive, tensile, twisting and bending forces that are set up as it transfers the to-and-fro motion of the piston into the rotary motion of the crankshaft. Why is the rod section usually of an **H** section as shown?

....

....

....

....

....

....

CONNECTING-ROD ALIGNMENT CHECKING

It is important that the axis of the big end bearing is parallel with the axis of the small end, within permitted limits, otherwise abnormal cylinder bore wear will result because of tilting of the piston.

Special jigs and fixtures may be used for checking this, or the connecting rod may be set up as shown opposite, that is, clamped to a mandrel mounted between centres in a lathe.

Note: The principles are the same whichever system is used.

INSPECTION

Check the alignment of four connecting rods and complete the table below.

What other faults may occur to connecting rods?

....

....

....

....

....

....

BEND   TWIST

CHECKING BEND

Mount dial gauge in toolpost and take readings at positions shown.
The difference is the amount of bend.

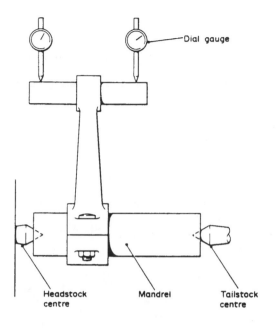

Dial gauge

Headstock centre   Mandrel   Tailstock centre

CHECKING TWIST

Indicate on the sketch below how to check for twist.

....

....

....

Rod No.	Bend		Twist		Recommendation
	Actual	**Allowable**	**Actual**	**Allowable**	

193

FOUR-CYLINDER IN-LINE CRANKSHAFTS

State the meaning of the following terms, as applied to crankshafts, and indicate one of each of them on the drawing below:

Main bearing journal ....

....

Crankpin journal ....

....

Throw ....

....

Web ....

....

Journal radius ....

....

Balance weight ....

....

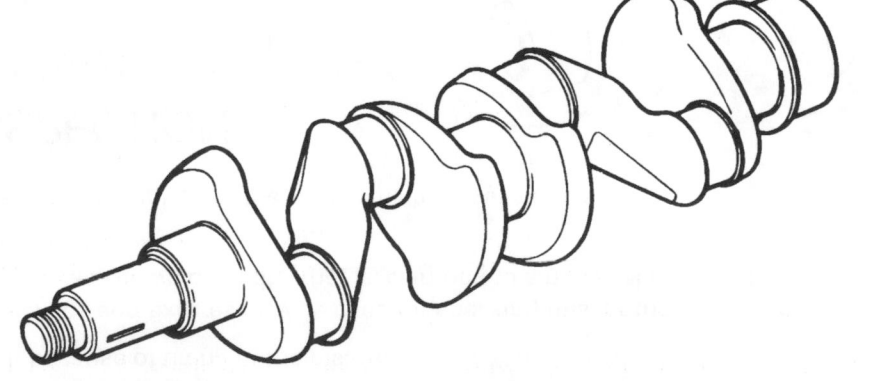

Type ....

Crankshafts are commonly made from drop forgings. What does this mean?

....

....

....

....

....

....

5-main bearing dropped forged crankshaft

© ROVER

An alternative method to produce crankshafts is by casting.

© FORD

3-main bearing cast crankshaft. The large webs can be hollow.

List the advantages of drop forged crankshafts compared with cast crankshafts:

....

....

....

....

Why is it considered desirable to use 5-main bearings instead of 3?

....

....

CYLINDER NUMBERING AND FIRING ORDER

There are two practicable firing orders for a four-cylinder in-line engine, these are:

.... and ....

With the aid of these firing orders, complete the tables below to show which one of the four strokes in the four-stroke cycle each individual cylinder is on at any given instant.

INSTRUCTIONS

1. Complete the firing orders, one in each table.

2. Fill in the table using the code:
   I, induction; C, compression; P, power; E, exhaust.

Cylinder number	Firing order 1		4	
1	P			
2				
3				
4				

Cylinder number	Firing order 1		4	
1	P			
2				
3				
4				

READING TABLE A

When 1 is on power stroke 2 will be on ....

When 4 is on exhaust stroke 3 will be on ....

READING TABLE B

When 3 is on induction stroke 2 will be on ....

When 1 is on compression stroke 4 will be on ....

Give an example of each of the two firing orders above:

Make	Model	Firing order

FOUR-CYLINDER VEE CRANKSHAFTS

The most common Vee-four engine arrangement in the UK uses the 60° Vee as shown.

Name a make and model of vehicle using this type of power unit.

Vehicle make .... Model ....

Show positions of cranks if engine is to be designed to have equal firing intervals.

Number the cylinders below and state a suitable firing order.

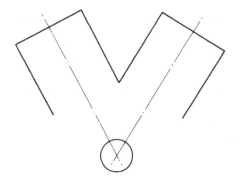
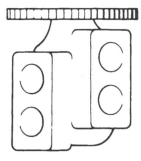

....

What unusual feature does this type of engine require with regard to engine balance?

....
....
....
....
....
....
....

SIX-CYLINDER IN-LINE CRANKSHAFTS

The diagrams below show the most common crank-throw arrangements.

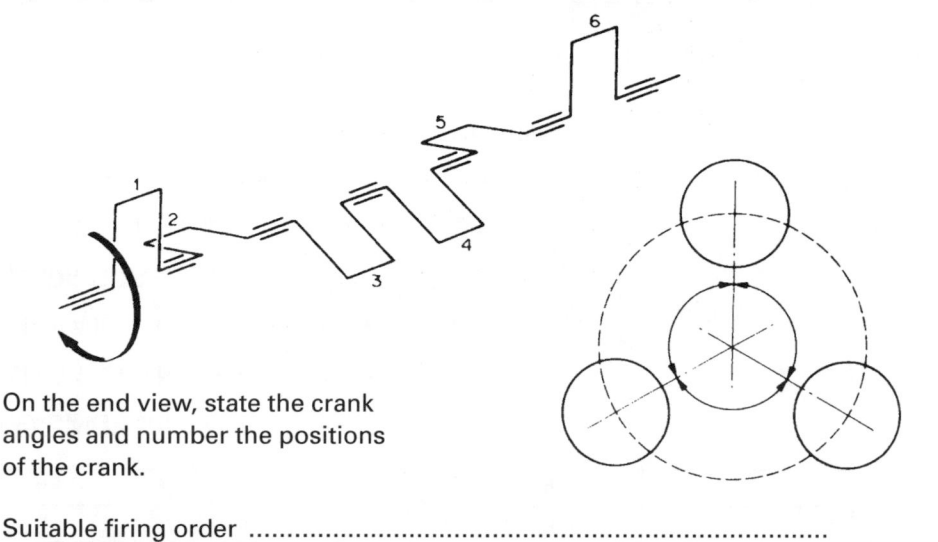

On the end view, state the crank angles and number the positions of the crank.

Suitable firing order ....

BUILT-UP CRANKSHAFTS

Some crankshafts are built-up in sections. What are the main advantages and disadvantages of this practice?

ADVANTAGES

....

....

....

....

DISADVANTAGES

....

....

....

Such a shaft is fitted to ....

SIX-CYLINDER VEE CRANKSHAFTS

Show positions of cranks if engine is to be designed to have equal firing intervals.

Number the cylinders below and state a suitable firing order.

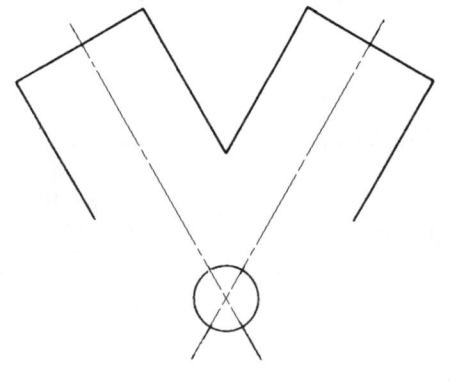

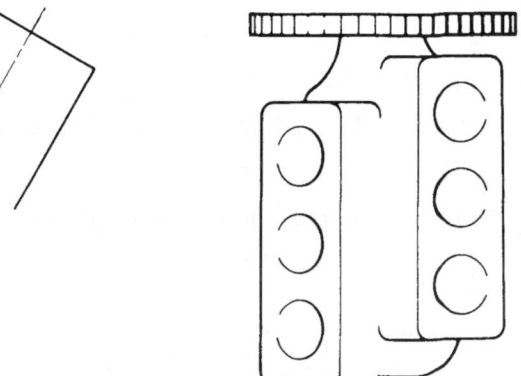

Firing order ....

Describe the cranks arrangement and its movement to create the firing order.

....

....

....

....

....

....

How would a Vee-12 crankshaft design differ from a Vee-6?

....

State the 'Jaguar Vee-12' firing order:

....

VEE-EIGHT CRANKSHAFTS

When the Vee is a 90° angle and the crankshaft designed as shown, the engine's primary and secondary forces can be perfectly balanced and the firing impulses evenly spaced.

The 90° angle, however, means that the engine is very wide.

If the angle is narrowed to 60° the crank becomes unbalanced and the firing intervals occur at unequal angles.

List some of the principal features of V8 engines.

....

....

....

....

....

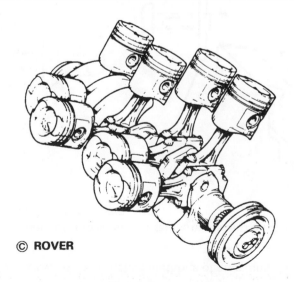

© ROVER

How many main bearings has the V8 CI engine shown? ....

Number the cylinders on the engine block.

Number the crankpins on both views of the crankshaft.

State a suitable firing order:

....

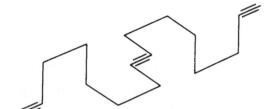

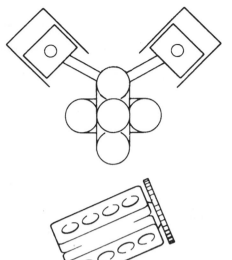

TWO-PLANE CRANKSHAFT

Show two alternative cylinder numbering methods, quoting vehicles and firing orders:

Make and model

....

firing order

....

Make and model

....

firing order

....

State the advantages of using Vee engines compared with in-line types:

....

....

FIVE CYLINDER CRANKSHAFTS

Five cylinder crankshafts may be used when insufficient space is available to fit a straight six engine.

State a five cylinder firing order ....

ATTACHMENT OF COMPONENTS TO THE CRANKSHAFT

The sketches below show two alternative methods of attaching the flywheel to the crankshaft.

They indicate dowel and locking washer arrangements and on the first diagram a method of oil sealing. Name the main parts.

The timing gear and crankshaft pulley are usually fitted at the front end of the crankshaft and the flywheel is fitted at the rear.

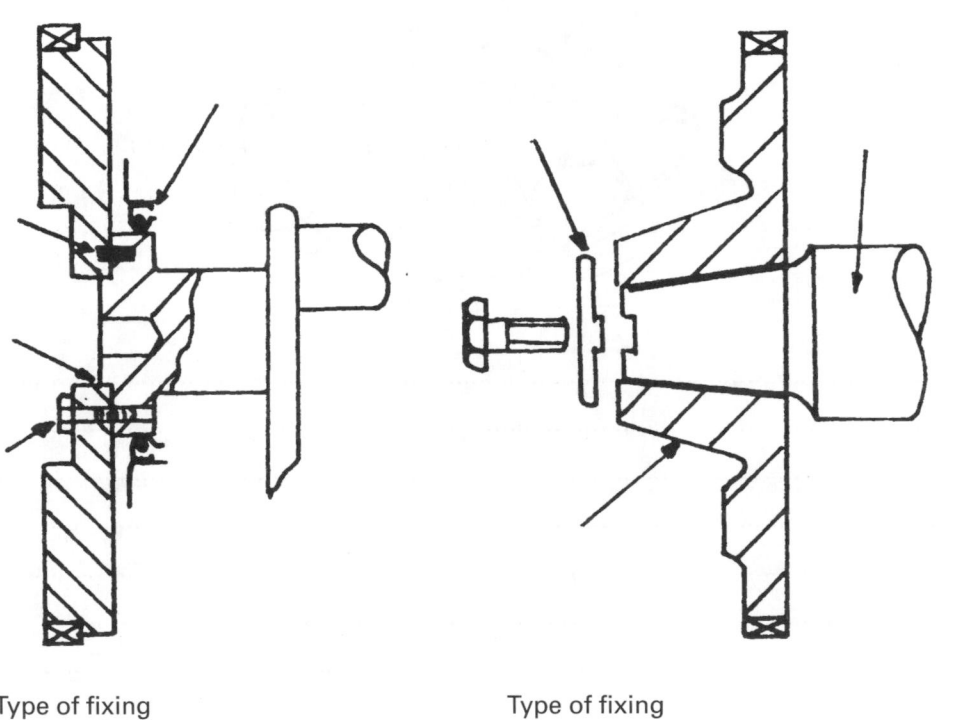

© **TALBOT**

List common faults found during service associated with:

1. crankshafts?

....

....

....

....

....

....

....

Type of fixing

Type of fixing

2. main and big end bearings?

....

....

What is the object of fitting dowels as well as studs to the flange type flywheel mounting?

3. flywheel?

....

....

....

....

CRANKSHAFT ALIGNMENT

After a period in service a crankshaft will become worn on its journals. This can be recognised and measured relatively easily.

Maximum wear is ....

....

Two other types of wear are commonly found on crankshaft journals. On each sketch below, indicate the shape (exaggerated) of one type of wear, name it and show where measurements would be taken.

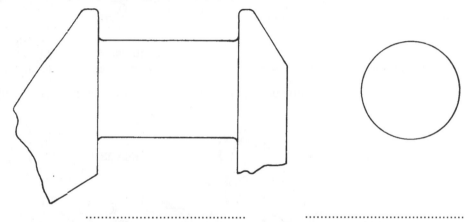

....

The crankshaft may also suffer from a less obvious defect such as becoming out-of-alignment due to twist and/or bend (or bow).

Title appropriately the two diagrams below:

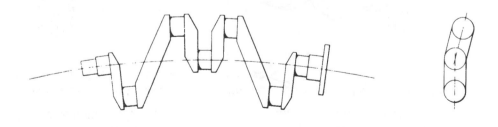

This shaft is ....   This shaft is ....

PRINCIPLES OF CHECKING CRANKSHAFT ALIGNMENT

Set up a crankshaft between Vee blocks as shown and check for bend and twist.

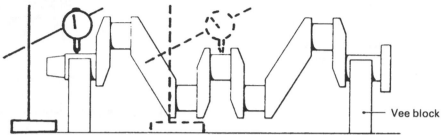

— Vee block

With the aid of the above diagram, explain how the shaft would be checked for bend:

....

....

....

....

....

....

....

The above drawing shows where readings should be taken to check the shaft for twist.

END-FLOAT – CRANKSHAFT

What is the function of the parts labelled 'A'?

....

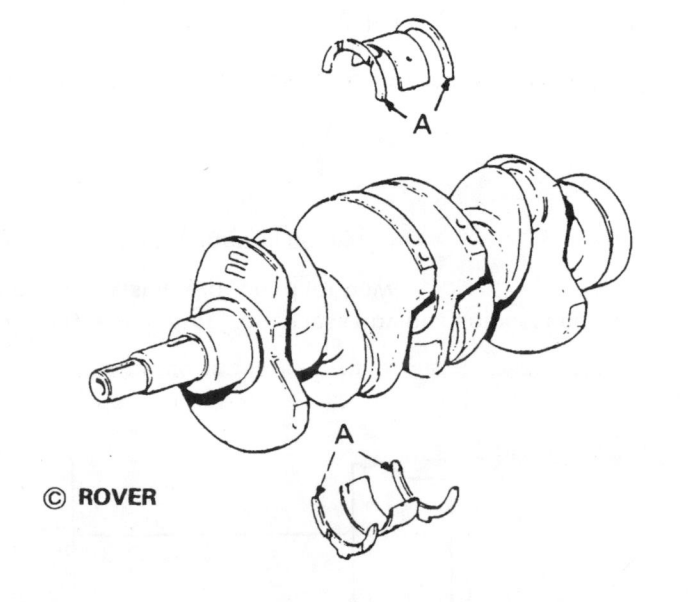

© ROVER

Crankshaft end-float should normally be between limits of ....

The clearance is checked using feeler gauges or by using:

....

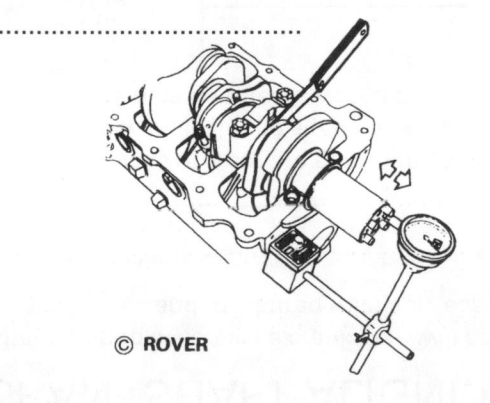

© ROVER

How can the end-float be altered?

....

....

The end-float clearance when measured as shown was 0.45 mm but the correct end-float should be 0.15 mm. The thickness of the thrust washers when removed was found to be 2.85 mm each. What should be the thickness of the new thrust washers?

....

INSPECTION

By means of using a strip of plastigauge material, or by any other suitable manner, determine the running clearance of a four-cylinder crankshaft's main and big end bearing journals

Engine make ....

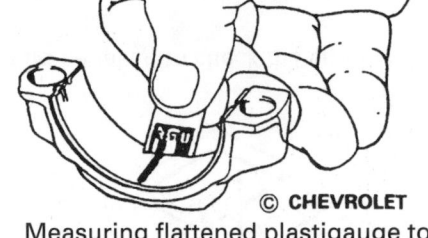

© CHEVROLET

Measuring flattened plastigauge to determine bearing clearance.

No. from front	1	2	3	4	5	Manufacturer's specification running clearance
Main bearings						
Big end bearings						

How is the plastigauge measured?

....

....

END-FLOAT – CAMSHAFT

In order to revolve freely at speed the camshaft must possess an end-float clearance. Typical limits are:..........................

The clearance may be checked as shown on the overhead camshaft arrangement below.

© ROVER ....

....

....

The correct camshaft end-float on four similar engines is 0.05 mm. Complete the table to show the extra thickness of pad required to bring the end-float to the recommended value.

Measured end-float	0.27	0.58	0.34	0.05
Extra thickness required				

ENGINE BALANCE

Balance weights are added to the crankshaft to balance the main inertia forces that would affect main bearing loading and engine smoothness. Examination of any crankshaft will show the positions of such balance weights; they are usually cast as part of the crankshaft. See page 194 – FOUR-CYLINDER CRANKSHAFTS.

What factors create the out-of-balance forces that require balancing?

....

....

....

....

In some engines (such as Vee-4 units) balance can only be achieved by fitting a reverse rotating balance shaft together with extra weights in the flywheel and crankshaft pulley.

Manufacturers of certain high-quality engines consider perfect balance to be essential and fit two balance shafts which rotate at double engine speed. These balance out the reciprocating forces which are known as secondary forces.

Describe the two shafts shown on the engines opposite:

....

....

....

How does the belt drive shaft arrangement differ from the chain drive type?

....

....

....

....

Vee-4 ENGINE

© FORD

Identify the position of the weights.

TWIN BALANCE SHAFT ENGINES

Describe with the aid of the sketches below how the balance shafts balance the forces created by the pistons and connecting rods.

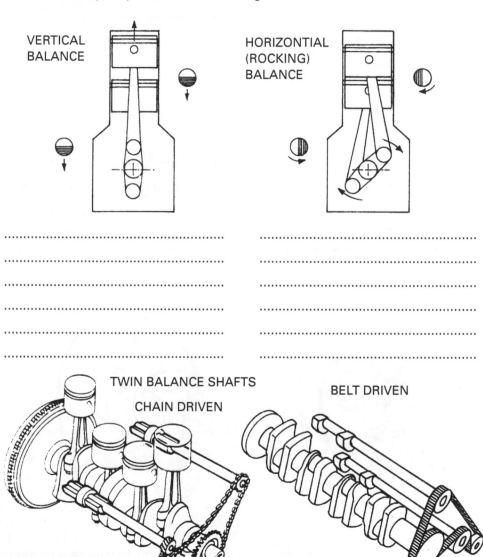

VERTICAL BALANCE

HORIZONTIAL (ROCKING) BALANCE

....

....

....

....

....

TWIN BALANCE SHAFTS

CHAIN DRIVEN

BELT DRIVEN

© MITSUBISHI   © PEUGEOT

CRANKSHAFT TORSIONAL VIBRATION

It is easy to realise that metal components vibrate with a certain natural frequency. Many factors contribute to the precise vibratory frequency, one of which is length.

This can easily be demonstrated by clamping one end of a short steel rule and flicking the free end.

Do the same with a longer steel rule and note the result:

The short steel rule had a .... natural vibratory frequency.

The long steel rule had a .... natural vibratory frequency.

State the causes of torsional vibration set up in a crankshaft:

....

....

....

....

....

....

Why is it important to stop vibrations from building up, even in a massive component such as a crankshaft?

....

....

....

....

CRANKSHAFT VIBRATION DAMPERS

Two common types of vibration dampers are shown below.

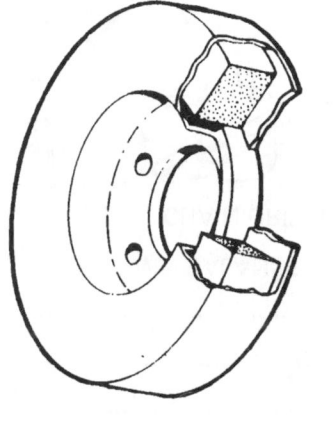 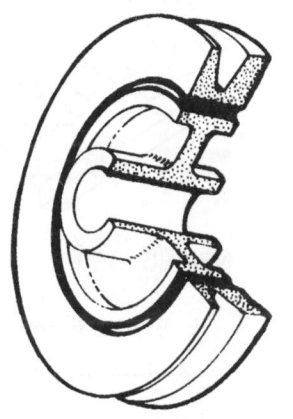

Type ....   Type ....

In the viscous damper, as vibration occurs the flywheel moves inside the casing shearing the silicone fluid. This requires a considerable force and it gives it the damping effect.

Describe the operation of the rubber type:

....

....

....

....

....

....

....

....

CRANKSHAFT VIBRATION FREQUENCY

The graph below shows the amount of oscillation build-up on a particular six-cylinder engine when the natural frequency of the crankshaft is at the same frequency as one of the 'critical' impulses of combustion.

Show on the graph the likely affect that the fitting of a rubber vibration damper would have on the amplitude of vibration.

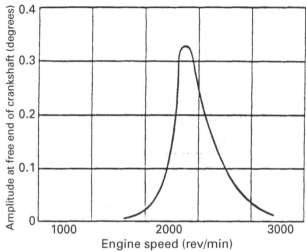

Describe how the damper controls crankshaft vibration.

....

....

....

....

....

List typical symptoms and causes of faulty engine vibration dampers:

Symptom	Cause
....	....
....	....
....	....

ENGINE MOUNTINGS

The engine/transmission unit is attached to the chassis or body structure through rubber blocks. Their main function is:

....

What are the *basic* causes of engine vibration?

....

....

....

....

Below is shown a typical front sub-frame / transverse engine mount layout.

Identify the engine mountings.

® HONDA

The mountings are considered to be fail safe. What is meant by this expression?

....

....

OVERHAULING/RECONDITIONING ENGINES

As part of your practical assessment you will have to be able to carry out efficiently many types of engine repair jobs. You have already described how to adjust valve clearances and fit engine timing belts.

State FOUR other items on which you expect to be assessed:

....

....

....

....

....

....

List TEN important points to consider when reassembling an engine after a complete overhaul:

1. ....

2. ....

3. ....

4. ....

5. ....

6. ....

7. ....

8. ....

9. ....

10. ....

....

What considerations should be given when selecting a replacement component for a given application?

....

....

....

....

MAIN STATUTORY REQUIREMENTS

List the main statutory requirements relating to engines under the following headings:

Silencers and noise

....

....

....

....

Exhaust and crankcase emissions

....

....

....

....

Power-to-weight ratios

....

....

....

Smoke emission ....

DIAGNOSTICS: ENGINE – SYMPTOMS, FAULTS AND CAUSES

State a likely cause for each symptom/system fault listed below. Each cause will suggest any corrective action required.

SYMPTOM	PROBABLE FAULT	LIKELY CAUSE
Drops of oil or coolant on garage floor after leaving overnight.	Oil and coolant leak	....
When starting from cold, engine cranks very slowly.	Too slow crank rotation	....
Regular heavy knocking particularly when starting from cold.	Bearing knock and rattle	....
The engine makes a rattling noise but only when cold.	Piston slap	....
Loud (explosive) banging noises from engine.	Backfire	....
A 'rattley' sound coming from the top of the engine.	Tappet rattle	....
The engine makes a 'pinking' noise when going up hills and carries on running when switched off.	Pre-ignition detonation pinking	....
Noticeable deterioration in engine performance.	Lack of power	....

DIAGNOSTICS: ENGINE (Continued)

State a likely cause for each symptom/system fault listed below. Each cause will suggest any corrective action required.

SYMPTOM	PROBABLE FAULT	LIKELY CAUSE
The engine keeps missing a beat when driving and runs roughly when the vehicle stops.	Misfire and uneven running	....
Exhaust emits blue smoke when the car accelerates.	Contaminated exhaust emission	....
Sump level needs topping up too frequently.	Heavy oil consumption	....
The engine is using far more fuel since the last service.	Heavy fuel consumption	....
The engine always requires a lot of cranking before it starts.	Poor starting	....
Temperature gauge reading is continually higher than normal.	Overheating	....
Smell of oily fumes and the air cleaner is full of oil.	Excessive crankcase blow-by	....
The dipstick level reading is rising and the oil seems thinner..	Crankcase dilution	....

ENGINE DIAGNOSIS – LACK OF PERFORMANCE

CYLINDER COMPRESSION TEST

The compression test is used to compare the pressure of individual cylinders. What would variations in pressure indicate?

....

....

EQUIPMENT

Describe test:

....

....

....

....

....

....

....

....

CYLINDER LEAKAGE TEST

The cylinder leakage test pressurises each cylinder in turn with compressed air, and if leakage is excessive checks are made to find from where it is escaping. The test clearly and quickly indicates the location of the compression leak and the seriousness of the leak in terms of cylinder's percentage compression loss.

State examples of leakage:

....

....

EQUIPMENT

Leakage Rates	
0–10%	Good
10–20%	Fair
20–30%	Poor
30% plus	Problem

Describe test:

....

....

....

....

....

....

....

....

ENGINE DIAGNOSIS – LACK OF PERFORMANCE

CYLINDER BALANCE TEST

With the engine running, each cylinder is 'shorted out' in turn and the variation in engine speed noted. How does this detect a fault?

....

....

....

....

EQUIPMENT   Engine analyser as shown opposite.

Describe test:

....

....

....

....

....

....

....

....

....

....

....

PRECAUTION

On emission controlled engines this test may not be recommended because unburnt fuel must not be allowed to reach the catalytic converter. Refer to car maker's instructions.

ENGINE PERFORMANCE TESTING USING CHASSIS DYNAMOMETER

© DYNO-TUNE/BUTTERFIELD

What types of checks and benefits can be carried out with the aid of a chassis dynamometer?

1. ....

2. ....

3. ....

4. ....

5. ....

6. ....

7. ....

8. ....

9. ....

10. ....

EXHAUST GAS ANALYSER

When fault-finding or tuning an engine, the use of an exhaust gas analyser will give a good indication as to whether or not correct combustion is being achieved.

There are several types of analysers, the most modern being the infra-red beam type. With this instrument an infra-red beam is passed across the flow of the gas and detects the amount of CO and in some types the amount of HC in the exhaust gas.

CO is ....

HC is ....

....

Sketch the type of exhaust gas analyser meter available for use in the workshop.

What other meter should be used in conjunction with the analyser?

....

State the tests that can be carried out with the meter:

....

....

....

....

Describe the general engine condition required before the meter is used:

....

....

....

....

....

Describe the meter setting-up procedure:

....

....

....

....

....

Check the exhaust gas content and performance of an engine. Connect an exhaust gas analyser and tachometer to the appropriate points of the engine.

Warm up the engine to its normal running temperature.

Use the analyser and complete the table below.

Vehicle make ....   Model ....

Type of carburettor/fuel injection equipment ....

Approximate testing speed	Actual values		Recommended values	
	CO	HC	CO	HC
Idle				
1000 rev/min				
2000 rev/min				

209

ENGINE PERFORMANCE TESTING

Below are shown two items of equipment used for measuring the torque and brake power of an engine.

The equipment is called ....

The types shown are ....

What is their basic function?

....
....
....
....
....
....
....
....

Spring balance

Tachometer

Loading weight

Sluice gate control wheel

Sluice plate

Water inlet

Rotor

Sluice gates

Casing

Sluice gate

Using an engine suitably connected to a dynamometer, obtain a series of brake load and fuel consumption values over the engine speed range and from these values calculate the torque, brake power and the specific fuel consumption. Use these values to construct graphs which will give an indication of the engine's outputs.

Engine make ....   Capacity ....

Describe the engine-testing procedure:

....
....
....
....
....
....
....
....
....
....
....
....
....
....
....
....

Engine power

The power measured by the dynamometer is known as the 'brake power'.

Define what is meant by brake power:

....
....
....

PERFORMANCE-MODIFYING PARTS

COMPRESSION RATIO

Give reasons why the compression ratio may be:

1. Raised ....

....

2. Reduced ....

....

How may compression ratios be:

Raised? ....

....

Lowered? ....

....

MODIFYING PISTONS

When would oversized pistons be fitted?

....

....

....

How would a set of oversized pistons be selected?

....

....

....

....

Examine engine data and quote piston sizes available for a particular engine.

Engine make .... Capacity ....

Original bore diameter ....

Available sizes of pistons ....

MODIFYING CAMSHAFTS

Why should altering valve timing or cam profiles be considered necessary?

....

....

How is the valve timing altered?

....

....

Sketch the profiles names.

Rapid rise high lift cam

Conventional cam

Conventional cam modified to high lift cam

CAMSHAFT RECONDITIONING

When would reconditioning a camshaft be considered necessary?

....

....

How is reconditioning achieved?

....

....

....

DISTORTED HEAD OR BLOCK

When does distortion of heads or blocks most commonly occur?

....

How are they reconditioned? ....

....

PERFORMANCE-MODIFYING PARTS

PRESSURE CHARGING

When is pressure charging required?

....

VALVE SEAT AND GUIDE RECONDITIONING

When should valve seats be reconditioned?

....

....

....

....

When should the valve guides be reconditioned?

....

....

....

How is a new valve seat insert fitted?

....

....

....

....

How is a new valve guide fitted?

....

....

....

....

CRANKSHAFT RECONDITIONING

When is it necessary to recondition a crankshaft?

....

....

....

How are crankshafts reconditioned?

....

....

....

SAFE WORKING PRACTICE

List some of the special precautions or general rules to be observed when testing, overhauling or repairing engines:

1. ....

....

2. ....

....

3. ....

....

4. ....

....

5. ....

....

6. ....

....

7. ....

....

Chapter 10

Engine Pressure Charging Systems

Pressure charging systems	214	Variable control turbochargers	220
Low-pressure supercharging	215	Turbochargers and interrelated components	221
High-pressure supercharging	215	Turbocharger testing	222
Mechanically driven pressure chargers	216	Diagnosing turbocharger faults	223
Turbocharger	218	Diagnostics	225
Boost pressure controls	219	Charger system protection during use	226
Intercoolers	220	Main statutory requirements	226

PRESSURE CHARGING SYSTEMS

The performance of an internal combustion petrol engine depends to a very large extent on the density or weight of the charge in the cylinder at the beginning of the compression stroke, that is, on the volumetric efficiency.

The volumetric efficiency of an engine and hence the power output can be improved by using a supercharger (or pressure charger) to force the charge into the cylinders during induction.

Pressure charging systems used on spark-ignition and compression-ignition engines can be of various designs. Briefly describe the TWO most common methods which are:

1. Exhaust gas driven turbochargers ....

....

....

....

2. Mechanically driven superchargers ....

....

....

Describe the two less familiar designs which are:

1. Pressure wave superchargers ....

....

....

2. Dynamic superchargers – rotating impeller ....

....

....

....

The dynamic supercharger is ideal for use on constant speed engines, e.g. aeroplanes. If used on road vehicles it produces too little pressure at low speeds and too much at high speeds.
The turbocharger's impeller design however, is very similar but it is much smaller and it is not mechanically driven.

State FOUR functional requirements of the pressure charging system:

1. ....

....

2. ....

....

3. ....

....

4. ....

....

The graphs below show typical power and torque curves for an unblown engine. Show by sketching on these graphs the effects of pressure charging for different levels of boost pressure.

Engine speed   Engine speed

Why is the pressure charging of CI engines much more popular than the pressure charging of petrol engines?

....

....

....

....

....

LOW-PRESSURE SUPERCHARGING

With low-pressure supercharging, air is blown into the cylinder to give an increased weight of charge and the engine compression ratio is lowered slightly. This increases the 'mean effective pressure' (mep) throughout the speed range, while still maintaining similar maximum compression pressure and maximum engine speed.

Why is it desirable to limit maximum compression pressure and maximum engine speed?

....

....

....

What is meant by the term 'mean effective pressure'?

....

....

....

HIGH-PRESSURE SUPERCHARGING

High-pressure supercharging is mainly limited to high-performance racing engines. The boost pressure is in the region of 2 bar (30 lbf/in^2) which increases the compression pressure and the maximum speed of the engine thus giving an overall increased performance.

What special features enable these engines to cope with the increased pressures and speeds?

....

....

....

State typical boost pressures and pressure charging ratios for a supercharged:

	BOOST PRESSURE	PRESSURE CHARGING RATIO
Conventional engine	....	....
Racing engine	....	....

Some superchargers are designed not to operate until the engine reaches a speed of about 2000 rev/min or the throttle is rapidly pressed to maximum.

How is this achieved? ....

....

....

How may increased pressures in the combustion chamber affect the combustion process in the:

(a) spark-ignition engine? ....

....

....

....

....

....

....

(b) compression-ignition engine? ....

....

....

....

....

....

Explain why it may be necessary to alter the compression ratio of an engine when installing a supercharger to an untuned engine.

....

....

....

MECHANICALLY DRIVEN PRESSURE CHARGERS

ROOTS SUPERCHARGER – ROTATING LOBE

Mixture is drawn from the carburettor and blown into the manifold at a boost pressure of 5.7 psi (35 kPa) on the Lancia design shown.
Indicate the lobes, note their direction of rotation and describe their shape.

....

....

....

....

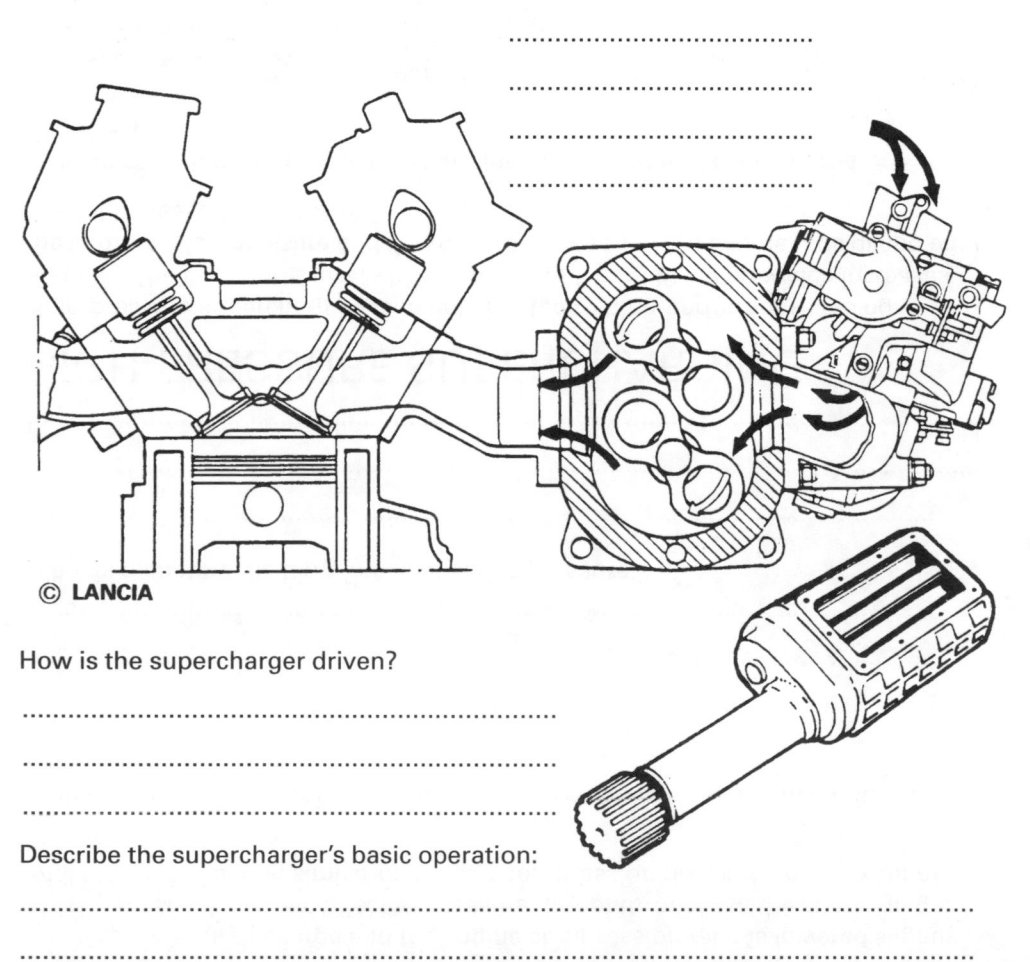

© **LANCIA**

How is the supercharger driven?

....

....

....

Describe the supercharger's basic operation:

....

....

....

SLIDING VANE

In the type shown an eccentric rotor and four vanes form compression spaces of varying volume. These chargers are capable of producing pressure up to 2 bar (200 kPa), and this gives a pressure ratio of 2:1.

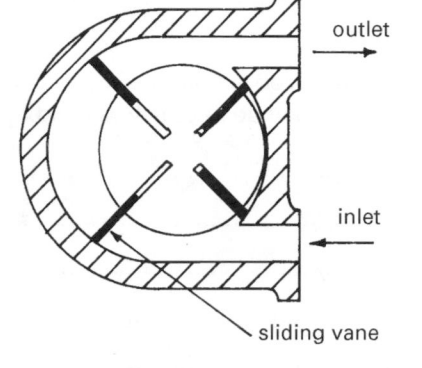

outlet

inlet

sliding vane

State advantages of the:

ROOTS TYPE

1. ....

2. ....

SLIDING VANE TYPE

1. ....

2. ....

On a petrol engine that uses a carburettor, where is the supercharger normally fitted?

....

Why is it fitted in this position?

....

....

....

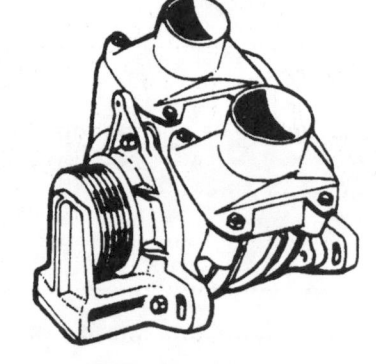

What type of belt provides the drive?

....

ROTARY PISTON OR SCREW COMPRESSOR

The Sprintex supercharger shown is driven by a toothed belt from the crankshaft.
Describe its operational cycle with the aid of the drawings opposite.

....
....
....
....
....
....
....
....
....
....
....

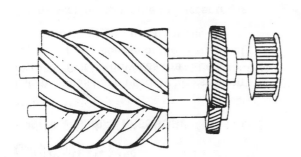

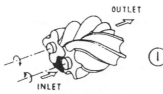

OUTLET

①

INLET

②

③

④

⑤

⑥

⑦

Name the parts on the sketch left.

PRESSURE WAVE SUPERCHARGER

A Comprex pressure wave supercharger is shown. This type is only suitable for compression-ignition engines, that is, displacing air only. It is mechanically driven but the pressure is provided from the exhaust gas. Describe its operational cycle with the aid of the drawing opposite.

....
....
....
....
....
....
....
....
....
....
....
....
....
....
....
....
....
....

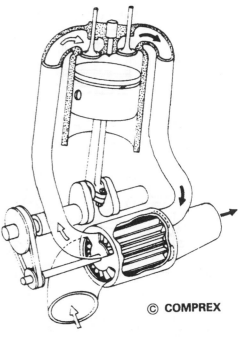

© COMPREX

On the sketch below, show the position of the supercharger.

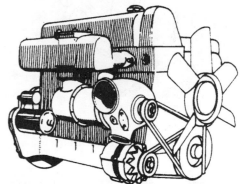

217

MECHANICALLY DRIVEN PRESSURE CHARGERS

ORBITING SPIRAL PRESSURE CHARGER

This type has been used by Volkswagen since 1986. It is known as the G-Lader supercharger because of the G shape of the spiral compression channels.

On the sketch shown the top eccentric is driven by a belt from the bottom eccentric at the same rotational speed.

Describe how the air is drawn through the spirals:

....

....

....

....

....

....

....

....

....

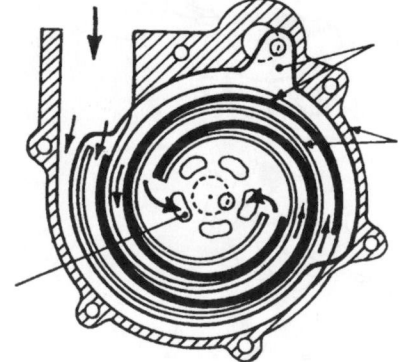

What are the advantages of this type of unit?

....

....

....

....

TURBOCHARGER

This type of pressure charger is driven by the flow of exhaust gas as it leaves the engine.

Name major parts and show the direction of rotation on the turbocharger shown.

Diagram shows relative size of rotor blades.

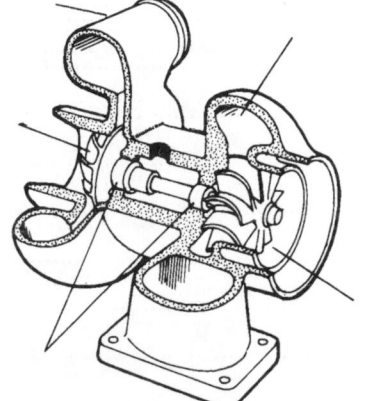

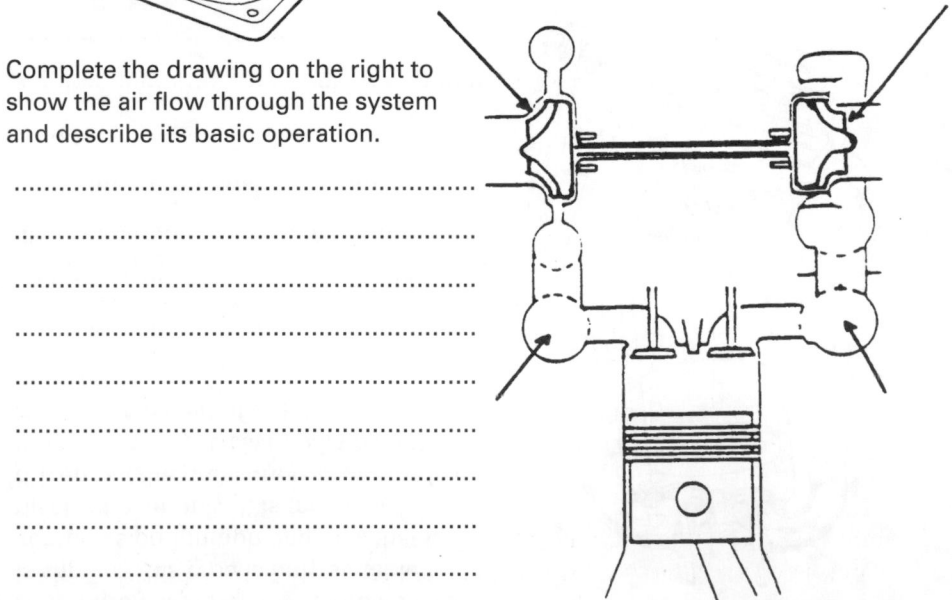

Complete the drawing on the right to show the air flow through the system and describe its basic operation.

....

....

....

....

....

....

....

....

....

....

BOOST PRESSURE CONTROLS

Unless controlled, turbochargers are capable of 'blowing up' engines at maximum speed.

State common types of control:

....

....

....

The most common method of control, particularly with turbochargers, is to redirect the air/gas through a waste gate.

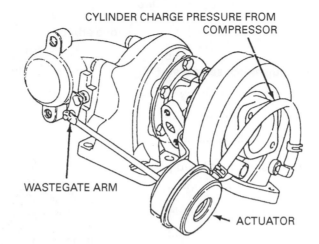

CYLINDER CHARGE PRESSURE FROM COMPRESSOR

WASTEGATE ARM

ACTUATOR

The layout below shows a turbocharger fitted to a petrol injection engine.
Indicate with arrows the direction of air flow through the system and name the main parts.

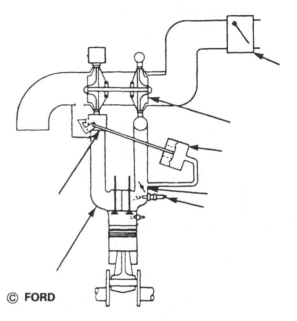

© FORD

FUEL INJECTION LAYOUT

Describe the operation of the pressure control valve shown. Name the parts indicated.

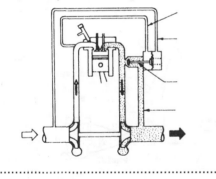

....

....

....

....

....

....

....

CARBURETTOR LAYOUT

Diagrams show possible positions of carburettor and blow-off valve or wastegate relative to turbocharger.

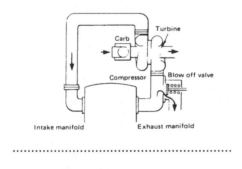

....

....

What are the advantages and disadvantages of this arrangement compared with mechanically driven superchargers?

ADVANTAGES ....

....

....

DISADVANTAGES ....

....

....

219

INTERCOOLERS

In a basic turbocharged system, the compressed air is forced directly into the engine as shown opposite.

On many engines an intercooler or charge cooler is used to cool the air after passing through the turbocharger (see also next page).

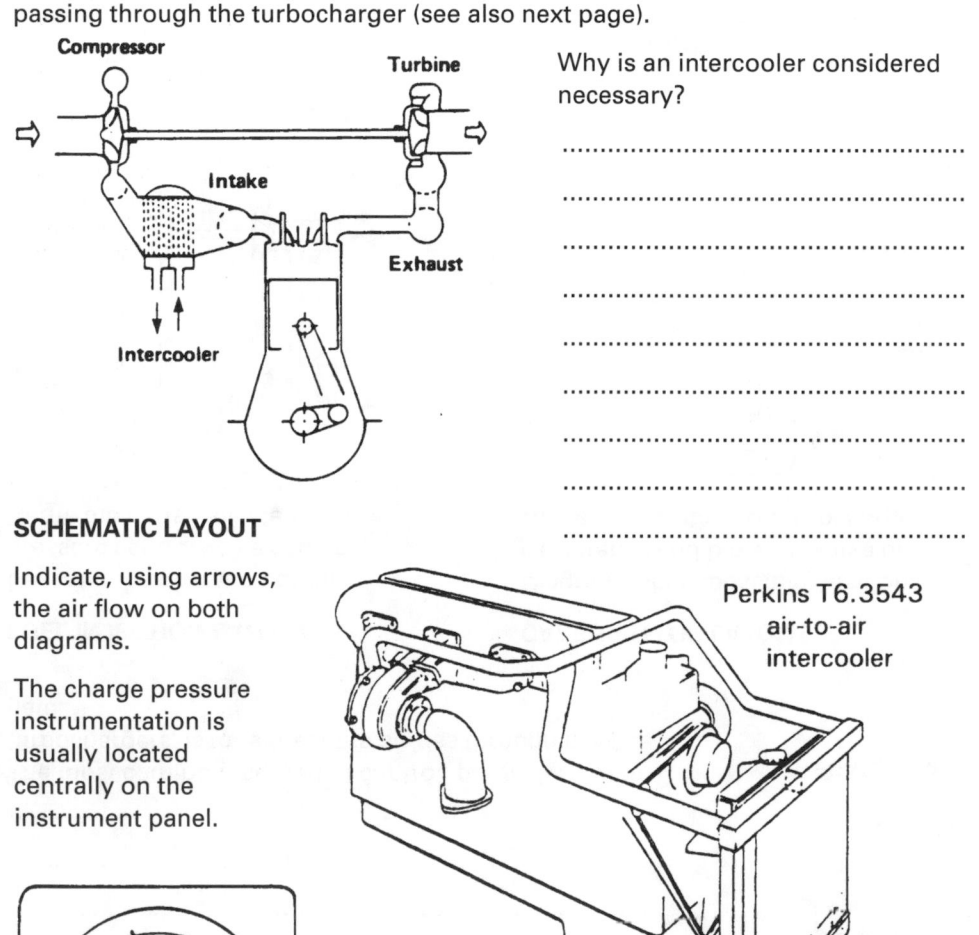

SCHEMATIC LAYOUT

Indicate, using arrows, the air flow on both diagrams.

The charge pressure instrumentation is usually located centrally on the instrument panel.

Perkins T6.3543
air-to-air
intercooler

Physical layout of engine and charge cooler

Why is an intercooler considered necessary?

....

....

....

....

....

....

....

....

VARIABLE CONTROL TURBOCHARGERS

To improve the output of turbochargers over a large varying engine-speed range, designs are being developed which contain a single, or many, movable flap(s).

Below is shown Nissan's NS-VN (variable nozzle turbo). The movable curved flap changes the turbo housing throat area in relation to engine requirements. The flap can move 27° and is operated by a vacuum diaphragm and pressure modulator from the engine ECU. Describe the turbine's action.

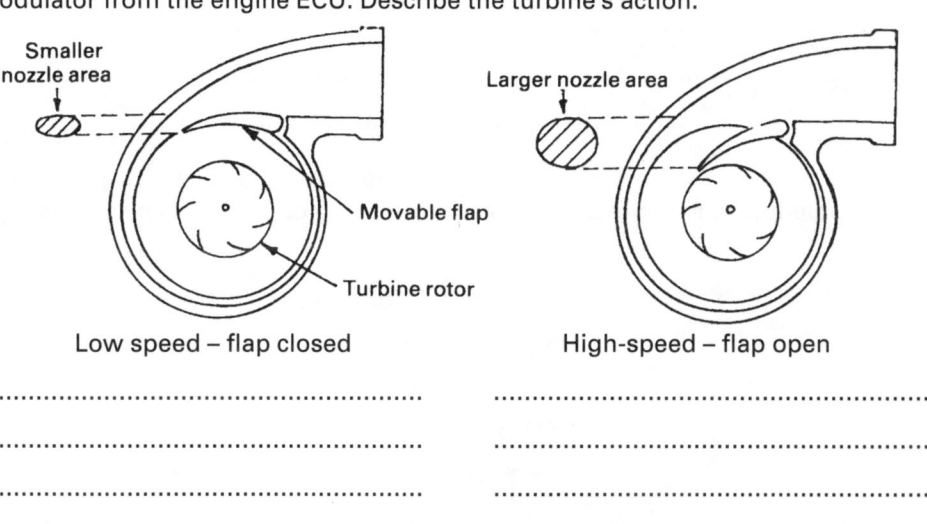

Low speed – flap closed   High-speed – flap open

....

....

....

....

....

....

What type of pressure will the above unit develop?

....

....

As well as increasing engine power output what other advantage should the flaps give?

....

TURBOCHARGER AND INTERRELATED COMPONENTS

Identify the components indicated.

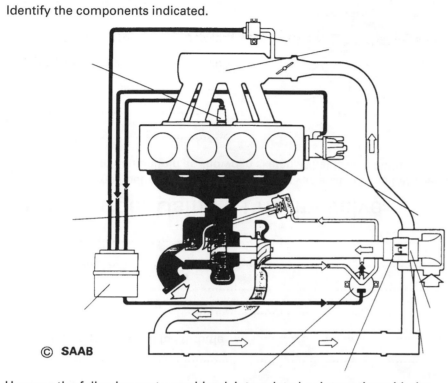

© **SAAB**

How are the following parts combined, interrelated or interactive with the pressure/turbo charger?

Exhaust manifold ....

....

Inlet manifold ....

....

Lubrication system ....

....

....

Plenum chamber (air box) ....

....

....

Engine cooling system ....

....

....

....

Air supply ....

....

....

Fuel supply ....

....

....

Engine management system ....

....

....

....

....

Engine combustion chamber, timing and bearings ....

....

Auxiliary drives ....

What possible speeds may turbochargers attain? ....

Turbocharger Operation

Describe the operation of a turbocharger at:

(a) Idling and lower part throttle ....

....

....

....

(b) Medium part throttle ....

....

....

....

(c) Upper part throttle and full throttle ....

....

....

....

Turbocharger Using a Priority Valve

Some turbocharged petrol engines use a 'priority valve'. It is fitted between the carburettor and inlet manifold as shown in sketch. What is its function?

....

....

....

....

....

....

TURBOCHARGER TESTING

SAFETY NOTE

Do not work on or around a turbocharger unless it is cold – turbochargers retain their heat for a long time.
Do not stick screwdrivers into the turbine. If a turbine blade is only slightly nicked it will cause imbalance at high speeds.

VISUAL INSPECTION – ENGINE STATIONARY
What external items should first be checked?

1. ....

2. ....

3. ....

4. ....

VISUAL INSPECTION – ENGINE RUNNING
With the engine running a normal turbocharger makes a slight whistling noise that increases in frequency with engine speed. When continually working on these vehicles a change from the normal sound should easily be detected.

What possible cause might the following types of noise indicate?

Louder than normal hissing noise:

....

Higher than normal pitch sound:

....

Uneven sound that changes pitch:

....

Grinding or rubbing sounds:

....

Uneven noise or vibration:

....

Rattling sound:

....

DIAGNOSING TURBOCHARGER FAULTS

LEAKS

Describe how the turbocharger should be checked for leaks with the engine running:

....

....

....

....

INSPECTING INSIDE OF TURBOCHARGER

Inspection may be possible on the vehicle or else the turbocharger may have to be removed. A torch or mirror may be required.
Describe what to look for when examining inside the unit:

1. Physical damage on blades ....

....

State the indicated faults on the blades shown below.

© BUICK

....

....

2. General damage on blades ....

....

3. Leakage ....

....

4. Wear on shaft ....
The fully floating bushes supporting the shaft have a clearance that may seem excessive when first examined. In general the clearances will be approximately:

Radial .... Axial ....

CHECKING BEARING CLEARANCES

After connecting dial test indicators as shown above describe how to check for:

1. Axial clearance	2. Radial clearance
....	....
....	....
....	....
....	....
CLEARANCE OBTAINED....................	CLEARANCE OBTAINED ....

3. Shaft for bend

....

....

....

Note. All these three different checks should be repeated several times to be sure of obtaining a correct reading.

State factors that would necessitate:

(a) repairing the unit ....

....

....

(b) fitting an exchange unit..

....

BOOST PRESSURE AND WASTEGATE OPERATION

Turbocharger boost and wastegate operation are usually directly related.
State the main cause of:

1. Insufficient boost pressure ....

....

2. Excessive boost pressure ....

ROAD TEST

The most accurate method of determining maximum boost is to check the
pressure during a road test.
Describe how such a test should be carried out:

....

....

....

....

....

....

If boost pressure is outside limits then an Actuator Operation and Linkage Test as
described opposite should be carried out.
What precautions must be observed when starting and stopping a turbocharged
engine?

....

....

....

....

....

WASTEGATE ACTUATOR AND LINKAGE TEST

To check the operation of the wastegate with the engine stopped or with the
turbocharger on the bench, a pressure tester (coolant radiator pressure tester
with suitable adapters) is ideal, and an accurate pressure gauge should be
connected in a similar way to that shown in the sketch below.

Describe the operation checks at the following stages:

1. Operation of actuator to check diaphragm for leaks

....

....

....

....

2. Movement of actuator rod

This movement is very small and a DTI is required to measure it.

....

....

....

....

3. Free movement of wastegate arm   Name items indicated.

© **VOLVO**

....

....

....

....

....

....

....

Note: Some wastegates can be adjusted, others must be replaced.

DIAGNOSTICS: PRESSURE CHARGING – SYMPTOMS, FAULTS AND CAUSES

State a likely cause for each symptom/system fault listed below. Each cause will suggest any corrective action required.

SYMPTOM/SYSTEM FAULTS	LIKELY CAUSE
1. After a sudden noise reduction there is smoking and oil leakage. 2. After parking there is water on the floor.	....
Large amounts of blue smoke issue from the engine when revving.	....
When the engine is revving an abnormal noise or vibration sets up.	....
When the engine accelerates it hesitates or lags excessively before it picks up speed.	....
Vehicle seems to be using more fuel than normal.	....
Engine will not produce sufficient acceleration and is low on power.	....
Poor starting and engine requires excessive cranking.	....
Engine lacks power and turbocharger smells excessively hot – components may have started to melt.	....

CHARGER SYSTEM PROTECTION DURING USE

Describe how the pressure charging system should be protected during use or repair from the following hazards:

1. Contamination ....

....

....

....

....

2. Overheating ....

....

....

....

3. Lubrication starvation ....

....

....

....

....

After oil and filter changing what is a recommended procedure before starting the engine?

....

....

....

How are seals and bearings protected to ensure long life?

....

....

MAIN STATUTORY REQUIREMENTS

List the main statutory regulations relating to pressure chargers under the following headings:

Exhaust gas content ....

....

Noise emission ....

....

Smoke and vapour ....

....

Power-to-weight ratio ....

....

Oil and fuel leakage ....

....

List the general rules/precautions to be observed to ensure efficient testing, overhauling and repairing of the pressure charging system when:

1. Working on vehicle after testing ....

....

2. Assembling chargers ....

....

....

3. Running and testing chargers ....

....

4. Running engines and using roller testers in confined spaces ....

....

5. Road testing ....

Chapter 11

Basic Electronics and Vehicle Electronics

Waveforms	228	Microcomputer components	239
Rectification	229	Engine management sensors	240
Capacitance	229	Manifold pressure sensor	241
Inductance	230	Electronic sensors	242
Transistors	230	Vehicle condition monitoring (VCM)	243
Resistors, colour coding and capacitors	231	Actuators	243
Electronic systems components – abbreviations	232	Testing equipment	245
Oscillator	233	Diagnosing vehicle sensor and circuit faults	245
Amplifier in ignition circuit	233	Automotive multimeter	246
Microprocessor control	233	System protection and precautions	246
Vehicle multiplexed control	233	Series circuits – effects of resistors	247
Electronic components (discrete)	234	Parallel circuits – effects of resistors	248
Electronic components – integrated circuits	236	Electrical testing problems	249
Drivers	238	Series circuits in electronic systems	250

WAVEFORMS

Identify the types of electrical waveforms shown:

....

....

....

....

....

A single sinusoidal wave is shown below. Label the items indicated.

(graph showing sinusoidal wave with m, 0.707m, 0.637m marked, and labelling lines)

....

Frequency = No. of rev/s

$$\text{Wavelength} = \frac{300 \times 10^6 \, \text{m/s}}{\text{Frequency}}$$

300×10^6 m/s = Velocity of light, which is the velocity of radio transmissions.

Identify the terms in the table by stating their quantity symbols and units.

TERM	QUANTITY SYMBOL	UNIT
Wavelength		
Frequency		
Period		
Amplitude		

Define the terms:

Cycle ....

Frequency ....

Hertz ....

Period ....

Amplitude ....

....

Wavelength ....

....

Complete the following relationships.

Period is inversely proportional to ....

The product of wavelength and .... is always the same. You may:

increase .... and reduce .... or

reduce .... and increase ....

The magnitude of waveforms is expressed in different ways. In the case of mains AC supply the magnitude is expressed as the rms value (root mean square value).

rms value = ...

Why is this rms value used?

....

....

....

....

....

....

228

RECTIFICATION

If you have completed Chapter 8 you will already be familiar with the rectification of an alternating current to direct current as produced in a vehicle's alternator. The alternator has a three-phase rectification system and this differs in many ways from the descriptions of rectification given on this page.

Rectification on this page is concerned with rectifying to a voltage suitable to operate electronic components and the diagrams shown are therefore in the style used by Electronic Engineers. Describe this style of wiring diagram.

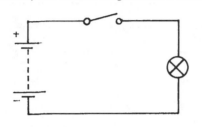

CONVENTIONAL DIAGRAM

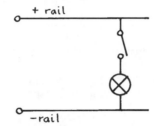

PREFERRED CIRCUIT DIAGRAM

....

....

....

....

....

HALF-WAVE RECTIFICATION

Sketch a typical circuit that would provide half-wave rectification and draw the voltage time graph of this rectification as would be shown on a CRO.

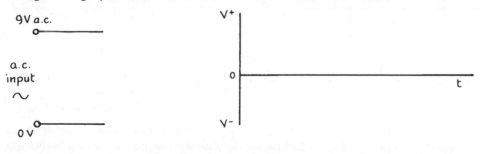

FULL-WAVE RECTIFICATION

Sketch a circuit to show full-wave rectification and the related time graph.

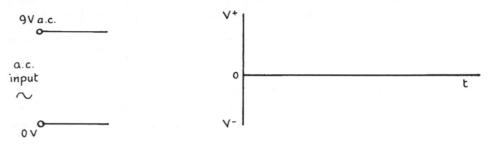

CAPACITANCE

What is the function of a capacitor?

....

Basically a capacitor consists of two conducting plates separated by an insulating material called a Dielectric layer.

The capacitor becomes electrically charged when a voltage is applied across it.

This charge is known as the

....

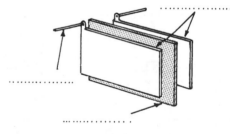

Basic construction of all capacitors

The maximum working voltage is dependent on the thickness and type of dielectric used. How is the capacitor affected if the maximum voltage is exceeded?

....

....

....

Capacitance is the property of the capacitor to store electrical charge when a potential difference is applied.

The unit of capacitance is...

INDUCTANCE

Whenever a circuit opposes a change in current (increasing or decreasing) the circuit is said to possess inductance.

What is an inductor?

....

....

....

....

....

....

MUTUAL INDUCTANCE

Describe how mutual induction is created in the right-hand coil when it is placed near the left-hand coil.

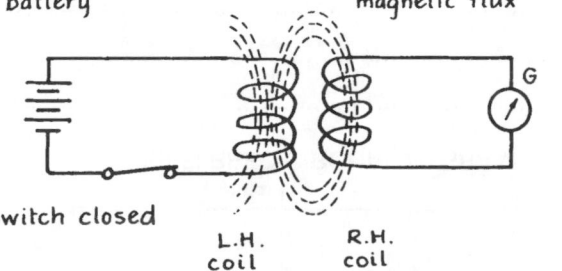

....

....

....

....

....

The most common example of an inductive circuit on a motor vehicle is the ....This type is said to have Mutual Inductance since a current change in one circuit induces an emf in another circuit.

TRANSISTORS

A transistor is similar to a diode with an extra side: one way it is a P – N – P sandwich; and positioned the other way it forms a N – P – N sandwich. Both types have three external connections.

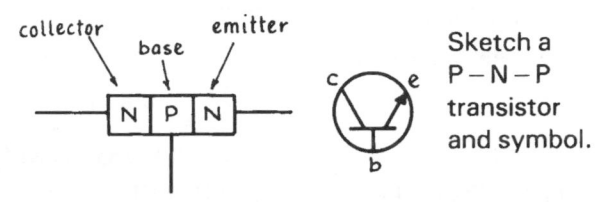

Sketch a P – N – P transistor and symbol.

TRANSISTOR ACTING AS A SWITCH

Describe the transistor operation to create current flow.

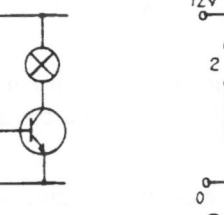 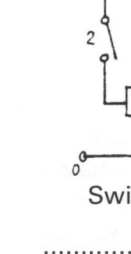 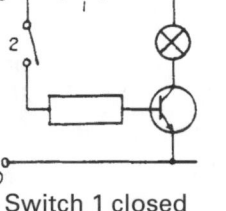 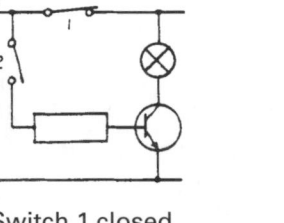

| Switches open | Switch 1 closed | Switches 1 and 2 closed |

....

....

....

....

TRANSISTOR ACTING AS AN AMPLIFIER

The amplification ratio of the collector to base current in the transistor is called the 'current gain'. A typical gain is 100. What is meant by transistor amplification?

....

....

....

....

....

RESISTORS

Describe the two main categories of resistors used in electronics:

1. Linear resistors

....

2. Non-linear resistors

....

....

....

COLOUR CODING

Small resistors have their value indicated by a colour-coded band system.

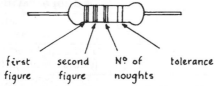

first figure second figure Nº of noughts tolerance

The resistor colour codes are as indicated in the table below.

Colour	Figure Ω	Tolerance (per cent)
Black	0	
Brown	1	±1
Red	2	±2
Orange	3	±3
Yellow	4	±4
Green	5	
Blue	6	
Violet	7	
Grey	8	
White	9	
Gold	–	±5
Silver	–	±10
None	–	±20

Identify the resistance of the coded resistors shown below:

1.

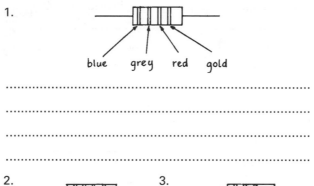

blue   grey   red   gold

....

....

....

2.   3.

orange black brown silver   brown green red

....

....

CAPACITORS

Electronic capacitors come in a range of many types. Identify the types shown.

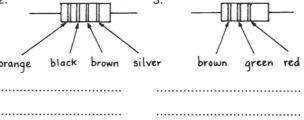

DISC TUBULAR

MICA   CERAMIC   POLYSTYRENE

Small capacitors are colour-coded in a manner similar to resistors.

CAPACITOR COLOUR CODE

Colour	Figure (pF)	Multiplier of Capacitance	Tolerance + –	DC Working Voltage
Black	0		20%	–
Brown	1	10	–	–
Red	2	100	–	250 V
Orange	3	1 000	–	–
Yellow	4	10 000	–	400 V
Green	5	100 000	–	–
Blue	6	–	–	630 V
Violet	7	–	–	–
Grey	8	–	–	–
White	9	–	10%	–

Identify the value of the following capacitors:

1.

red black green white red

....

....

....

2.

yellow violet orange black blue

....

....

3. What colour code will be on a capacitor having the following value:

1.0 µF 400 V working 10%?

....

....

ELECTRONIC SYSTEMS – COMPONENTS – ABBREVIATIONS

Electronics deals with the behaviour and effects of small amounts of electrons in semiconductor circuits. Electrical equipment uses larger amounts of current.

Name TWO devices that would be considered:

(a) Electronic

    1. ....

    2. ....

(b) Electrical

    1. ....

    2. ....

State a vehicle circuit or component which would incorporate electronic control or actions listed below.

CONTROL/ACTIONS	COMPONENT
Rectification	
Wrong polarity protection	
Regulation of power supply voltage	
An amplification signal using a Darlington pair	
An AC waveform oscillator	
A square wave form (multi-vibrator)	
Logic switching	
Microprocessor control	
Multiplexed control	

State the function of the following components which may be found in the charging system:

Rectifier ....

Reverse polarity relay ....

....

Voltage regulator ....

....

(Further details of these components may be found in other chapters.)

DARLINGTON PAIR

This is a pair of transistors which are connected together in such a way that the first transistor switches on the second transistor. This allows the very small current (0.1 mA), which triggers the first transistor, to be amplified by the 'gain' of the two transistors. It can give an amplification of up to 8000.

Give examples of the use of Darlington pairs.

....

....

....

....

....

....

....

Circuit showing a Darlington pair

This chapter contains numerous abbreviations relating to electronics. State the meaning of those listed below.

LED	FET	CPU ....
LDR	TTL	A/D ....
LCD	CMOS	D/A ....
CRT	RAM	F/V ....
CRO	ROM	PCB ....
VFD	EPROM	VCM ....

State the function of the following electronic components.

OSCILLATOR

The oscillator produces its own high-frequency signal, 3 to 4 MHz, whenever a voltage is produced across the sensing coils.

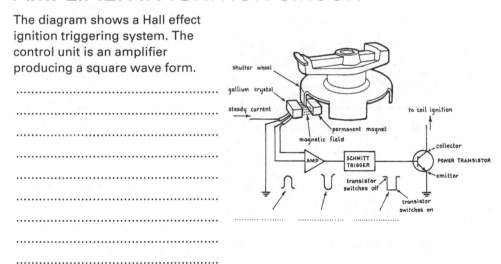

Speedo drive quenched oscillator speed sensor.

Output signal before and after filtering

....
....
....
....
....
....
....
....
....

AMPLIFIER IN IGNITION CIRCUIT

The diagram shows a Hall effect ignition triggering system. The control unit is an amplifier producing a square wave form.

....
....
....
....
....
....
....
....

MICROPROCESSOR CONTROL

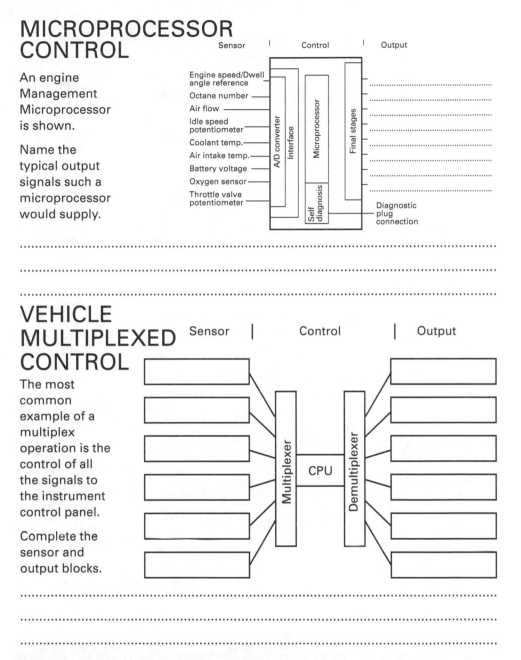

An engine Management Microprocessor is shown.

Name the typical output signals such a microprocessor would supply.

....

....

VEHICLE MULTIPLEXED CONTROL

The most common example of a multiplex operation is the control of all the signals to the instrument control panel.

Complete the sensor and output blocks.

....

....

233

ELECTRONIC COMPONENTS (DISCRETE)

Recognise the components shown, identify their wiring symbol and state their function or effect on design.

COMPONENT/SYMBOL	FUNCTION

COMPONENT/SYMBOL	FUNCTION

DISPLAYS

The multiplexed CPU shown earlier in this chapter gives an indication of how the display values are controlled.

Identify the items indicated.

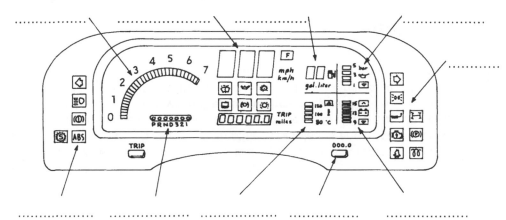

Displays may be considered to be 'active' or 'passive'.

What is meant by these terms?

....

....

....

Name FOUR types of electronic indicating displays:

1. ....

....

2. ....

....

3. ....

....

4. ....

....

CHOKES OR INDUCTORS

These are coils of wire usually wrapped round a former having an iron core. Describe their action.

Identify the wiring symbols:

....

....

....

1. ....

2. ....

3. ....

....

TRANSFORMERS

A transformer relies on electromagnetic induction to generate an emf in the secondary winding.

Altering the ratio of turns on the primary and secondary windings allows designs that can be used for increasing (step up) or decreasing (step down) voltages.

The magnetic field produced by the alternating current cuts through the secondary coil windings and produces and alternating e.m.f. in the secondary winding.

Identify the items indicated.

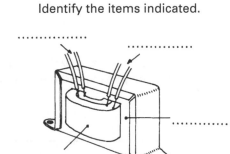

PRINTED CIRCUIT BOARDS (PCB)

These boards are used to simplify circuit arrangements. They replace the use of many interconnecting cables. Where may PCBs be used and what is the conducting material?

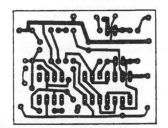

....

....

....

....

....

ELECTRONIC COMPONENTS

This and the next two pages are concerned with giving you an awareness of the types of semiconductor accessories that are available to build circuits. They may all be classed under the broad title of Integrated Circuits (IC).

INTEGRATED CIRCUIT

The integrated circuit is built up on a chip of silicon. The chip is usually mounted in a black plastic block having connecting lugs positioned on each side (DIL, dual in-line). The size of the block is determined by the number of connections required. The unit shown is a 14 pin DIL package.

Examine a chip used in such a unit and describe its assembly:

....

....

....

....

....

....

....

CIRCUITS TO PROVIDE BASIC FUNCTIONS

Simple integrated circuits have been developed which incorporate all the functions of items such as timers, amplifiers, regulators and multi-switches. These items must be considered as single components with inputs and outputs. It is only necessary to know what they do, not how they work. They are cheap to purchase and are used to build up electronic switching circuits on vehicles and any item that requires electronic control.

TRANSISTOR ARRAYS

....

....

....

....

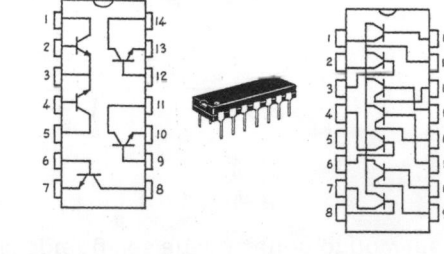

OPERATIONAL AMPLIFIER

The 741 is an integrated circuit on a silicon chip consisting of 20 transistors, 11 resistors and 1 capacitor.

The symbol for the op-amp is a triangle.

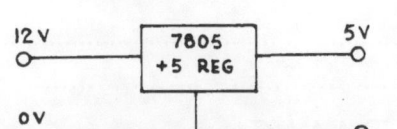

741 SERIES

OFFSET NULL 1 — COMP 8
INV INPUT 2 — V+ 7
NON-INV INPUT 3 — OUTPUT 6
V- 4 — OFFSET NULL 5

TOP VIEW

....

....

....

....

....

In the block diagrams draw graphs which show the typical use of an amplifier, i.e. where the voltage signal is amplified exactly to give a reading at the display.

sensor — amplifier — display

REGULATORS

These chips control the voltage at precise levels. The 7805 regulator shown has a 1 A output operating at 5 V.

....

....

12 V — 7805 +5 REG — 5 V

0 V

236

TIMERS

The 555 timer can be fitted into any circuit that requires: (1) to operate for a specific length of time (monostable); (2) to allow switching between two voltages, such as 12 V and O V continuously (astable). Also known as multivibrators.

Name circuits that may fit electronic timers.

MONOSTABLE   ASTABLE

....

....

....

....

Describe how the timer acts as a monostable unit.

....

....

....

....

....

MONOSTABLE TIMER OPERATING A COLD START HEATER RELAY

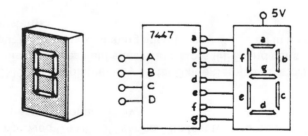

Describe the action of the timer control.

....

....

....

....

CONVERTERS AND DECODERS

An analogue-to-digital converter (A/D or ADC) is used on most electronic display circuits using numbers. What is such a converter's function?

....

....

Complete the graphs showing engine temperature to gauge A/D values.

Temp. as a resistance

Temp. as a number

Time

Sensor   A D C   85

A digital-to-analogue converter (D/A) may be required if the values have been multiplexed and require to go to analogue gauges.

SEVEN SEGMENT DISPLAY

A separate seven segment control is an IC, called a BCD to 7 segment decoder. The digital signals are converted into binary numbers, and to form the numbers certain LEDs must be lit.

Which circuits must be operated to give the following numbers?

2. ....

3. ....

4. ....

FREQUENCY-TO-VOLTAGE CONVERTER

These convert input frequencies into analogue voltages, the frequency being proportional to the voltage. They can also be used in the opposite mode – voltage to frequency.

DRIVERS

These are integrated circuits which offer many functions on one chip and so simplify circuitry while providing more driving power.

DISPLAY DRIVERS

These are decoder drivers designed to convert analogue input voltages directly to drive a common seven segment LED display or up to a ten-line bar display.

Describe the function of the following drivers:

STEPPER MOTOR DRIVERS..

....

....

DARLINGTON DRIVERS ....

....

....

....

LOGIC FAMILIES

Logic systems that have inputs and outputs make use of AND, OR and NOT gates to control mechanical input switches. What are logic families?

....

....

....

....

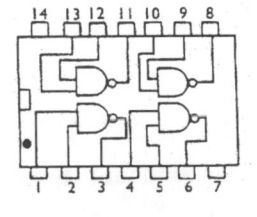  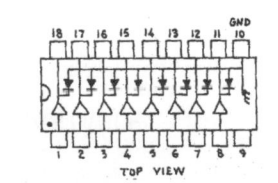

....

....

There are two different methods of producing electronic family logic chips; each uses a different type of transistor. The two families are known as:

TTL (Transistor Transistor Logic)
This type uses bipolar transistors and the ICs are listed as 7400 series or the improved '74 HCT00' series. The last two numbers refer to the type of gate. (Operating voltage is 5V)

CMOS (Complementary Metal Oxide Semiconductor)
This type uses field effect transistors (FET) and consumes very little current 8µA, but has a slow switching speed. They are fitted in electronic instrumentation systems.

A NAND gate is an AND gate connected to a NOT gate. It is possible to build all other gates from NAND gates.
This means that only one type of integrated circuit need be manufactured.

Describe what electronic components the diagram shows.

[diagram: 5V supply, TTL gate with 0.8 mA, Logic '1', 6k8 resistor, transistors, lamp, 0V rail]

....

....

MEMORY

One very important aspect of logic gates is the fact that they can be assembled in such a way that they can remember logic values even after these values have changed.

MICROCOMPUTER COMPONENTS

The microcomputers used on vehicles have fixed programmable memories which consist of sets of operating instructions provided by the manufacturer. All additional information is given to the microcomputer by the various sensors. This information is received, compared with the fixed data and, if the values are not similar, the microcomputer alters the engine's running until the values compare.

The main components of the microcomputer are shown in block form below.

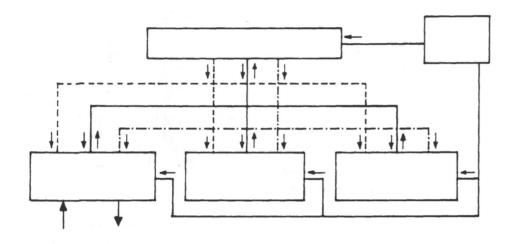

Describe the function of the Microcomputer parts named below:

CPU CENTRAL PROCESSING UNIT ....

....

....

....

ROM READ ONLY MEMORY..

....

....

....

BASIC FUNCTION OF MICROCOMPUTER PARTS

RAM   RANDOM ACCESS MEMORY   This stores information received from the sensors and is changed on each update. When the power supply is switched off all information in this memory is lost.

EPROM   ERASABLE PROGRAMMABLE READ ONLY MEMORY
This is a chip that can be programmed with new information and then used as a ROM chip. E.G. racing cars use EPROMs to change engine running characteristics for different types of tracks and weather conditions.

INPUT INTERFACE   Allows input of system data from sensors and manual controls to the CPU. It will incorporate an A/D convertor.

OUTPUT INTERFACE   Provides the connections via a D/A converter to the operating units which the CPU controls.

BUSES   These are three channels inside the computer along which the data flows, i.e. there are three sets of signals required to carry out each process.

(1) The data must be found in the memory at its 'address'

(2) and instruction where to send data must be given to 'control' and then

(3) the data must be moved.

CLOCK SIGNAL   Movement of data is controlled by at time pulse from a quartz crystal operating at a frequency of over 2 MHz. When a voltage pulse of the clock is applied simultaneously to two parts of the computer, then data will pass between these two parts along the bus. This is like high frequency multi-plexing.

CONTROL
UNIT FOR
ELECTRONIC
PETROL
INJECTION

Motronic M 3.3.1

(digital)

© BOSCH

SENSORS

Regardless of type, sensors indicate electronically a set of values which, if outside pre-set limits, must be corrected. This is done in normal running by the ECUs to provide efficient operation of the engine, transmission, brakes, suspension etc.; or it is corrected by the driver/mechanic if a warning indicator operates to signal a fault.

ENGINE MANAGEMENT SENSORS

Describe the action of the sensors shown:

1. ENGINE KNOCK SENSOR

....
....
....
....

© BOSCH

2. ENGINE SPEED and POSITION SENSOR

© BOSCH

....
....
....
....
....

3. ENGINE COOLANT TEMPERATURE SENSOR

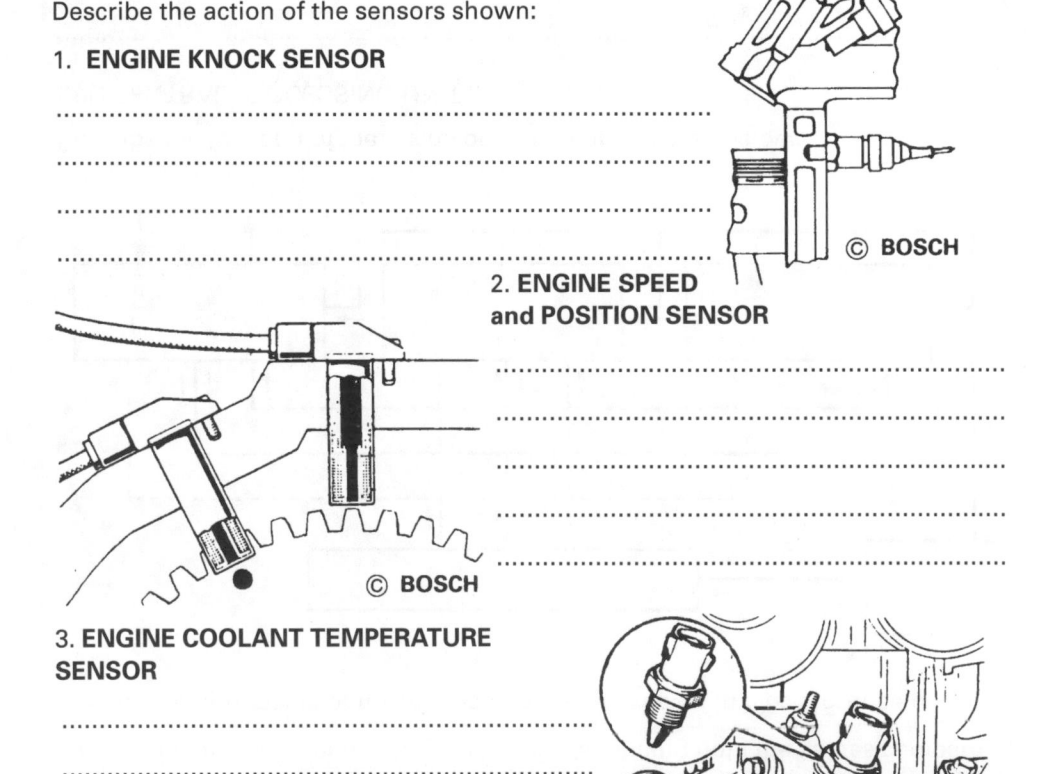

....
....
....
....

© FORD

4. INLET AIR TEMPERATURE SENSOR

....
....
....
....
....

© BOSCH

5. AIR FLOW METERS

(i) Vane Type

....
....
....
....

© FORD

(ii) Hot Wire Type

....
....
....
....
....
....
....
....

Sensing wires

© LUCAS

(iii) Optical Karman Vortex Air Flow Meter

This air flow meter, known as the KARMAN VORTEX type, measures the air flow volume optically. There is a pillar called the 'vortex generator' placed in the middle of the uniform flow of air, its obstruction causes the air to generate vortexes (Karman vortexes) down-stream of the pillar. These vortexes are measured and from the values obtained the air flow volume is determined.

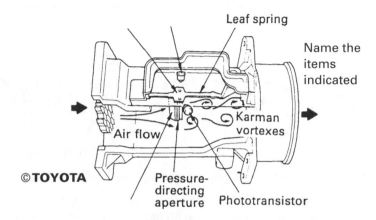

Name the items indicated

©TOYOTA

With the aid of the drawing right, describe how the vortexes are measured.

....

....

....

....

....

....

....

....

....

....

VIBRATING MIRROR

What advantages does the Karman vortex air flow meter have compared to the vane type? ....

....

MANIFOLD PRESSURE SENSOR

or MANIFOLD ABSOLUTE PRESSURE (MAP) SENSOR

The sensor shown measures the strength of vacuum in the intake manifold and so senses the volume of air by density. The ECU using this value, together with values from the throttle opening sensor and engine temperature sensor can determine the basic injection duration and basic ignition advance. No air flow meter is required in systems where this sensor is used.

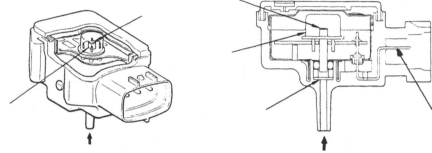

As shown in the diagram the sensor unit consists of a silicon chip combined with a vacuum chamber. One side of the chip is exposed to intake manifold pressure and the other side is exposed to the internal vacuum chamber. The engine ECU supplies a constant power source of 5V.

Describe how the intake manifold pressure is converted into a voltage signal.

....

....

....

....

....

....

....

....

GRAPH SHOWS HOW INTAKE MANIFOLD PRESSURE IS PROPORTIONAL TO OUTPUT SIGNAL VOLTAGE

ELECTRONIC SENSORS

PRESSURE CONTROL

State THREE items for which measurement of pressure is obtained.

....

Name types of electric pressure sensors:

....

....

....

....

....

All these sensors require pressure to create movement which can be converted into an electrical voltage or resistance that is proportional to the pressure measured, be it large or small.

LEVEL SENSORS (Note: For engine oil level see page 27.)

The level of fuel in the tank is indicated by the position of the float; this operates a transmitter using a thick film resistive track and wiper.

With the aid of the diagrams, describe how a low fuel warning will be given.

Name the parts indicated.

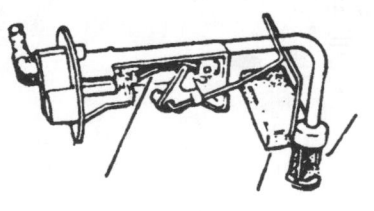 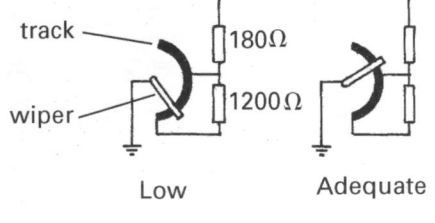

Low   Adequate

....

....

....

....

Coolant and windscreen washer levels are operated by reed switches controlled by permanent magnets. Describe the action of the reed switch.

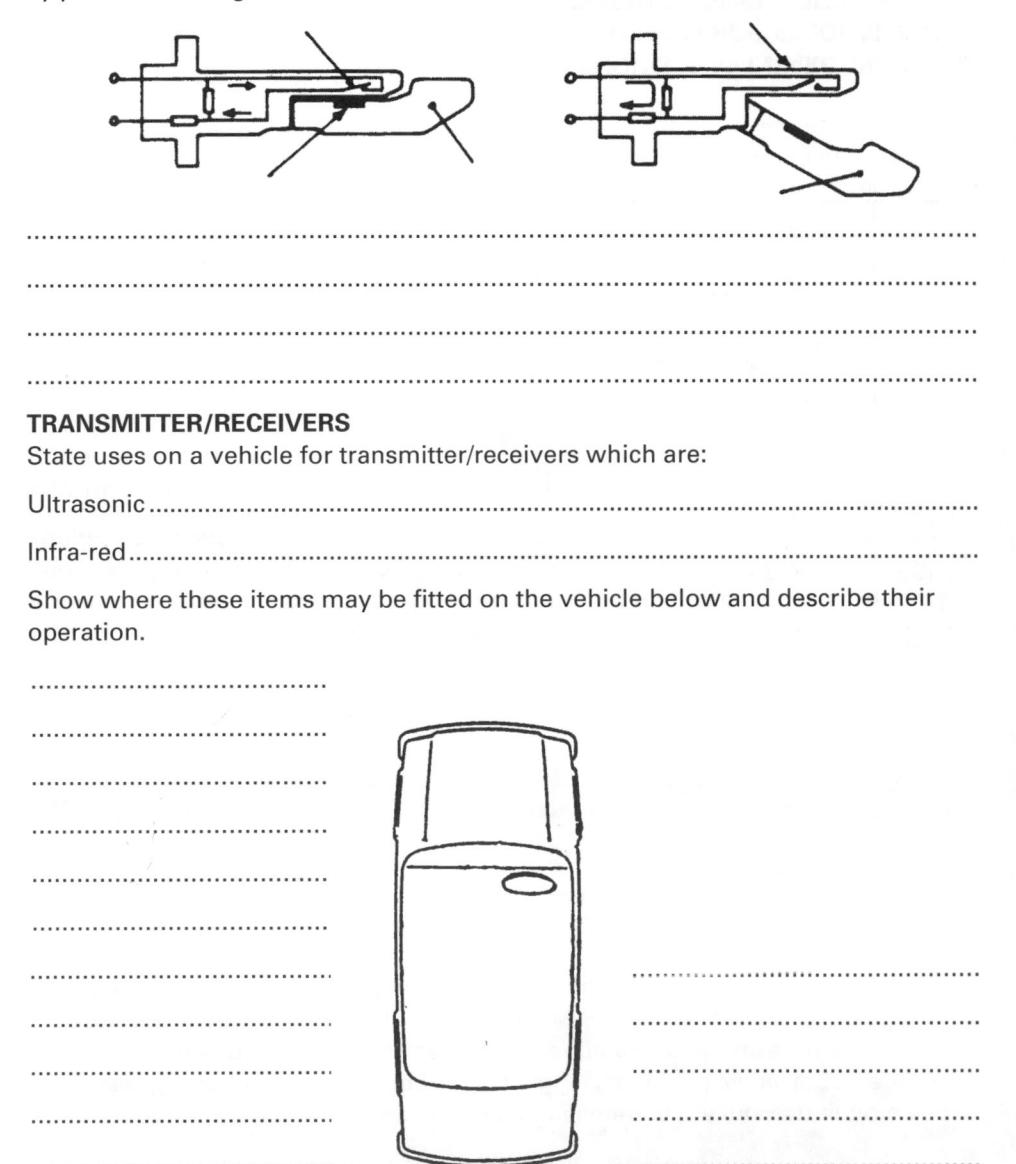

....

....

....

....

TRANSMITTER/RECEIVERS

State uses on a vehicle for transmitter/receivers which are:

Ultrasonic ....

Infra-red ....

Show where these items may be fitted on the vehicle below and describe their operation.

....

....

....

....

....

....

....

....

....

VEHICLE CONDITION MONITORING (VCM)

This provides the driver with an indication of the operational state of the systems that are vital to the safety and operation of the vehicle, or are subject to legislation.

Name systems/functions that are generally checked using Vehicle Condition Monitoring.

....

....

....

....

SENSORS FOR LIGHTING SYSTEMS

How is a lamp failure or open circuit indicated?

....

....

....

....

Lighting and door-open faults are often indicated on a vehicle map, similar to the one shown opposite.

Examine a vehicle equipped with such a map and state the colours that will appear when a fault occurs.

Main and dipped beam ....

Side, tail and no. plate ....

Rear fog and brake ....

Turn indicators ....

Doors open ....

Name or colour the indications

ACTUATORS

Actuators, when given an electrical signal, produce mechanical motion which (depending on design) is either rotary or linear.

SOLENOIDS AND RELAYS
On what electrical effect do both these components rely?

....

....

....

....

A relay is commonly activated by a transistorised switch and then passes a large operating current (see page 150). The most common solenoid is the linear type which switches and operates the starter motor (see page 154).

Name two other solenoid controls: ....

....

DOUBLE ACTING SOLENOID
A double acting solenoid is shown in principle below.
Describe its operation and state where it may be used.

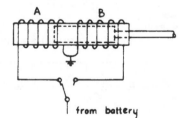

from battery

ROTARY SOLENOID
This unit provides a rotary movement of 45°. A return spring and stop are fitted at the rear end of the shaft.

....

....

....

....

STEPPER MOTORS (see also page 261)

Stepper motors are designed specifically to control movement accurately. How can the motor move in a series of steps each turning the motor 7.5°?

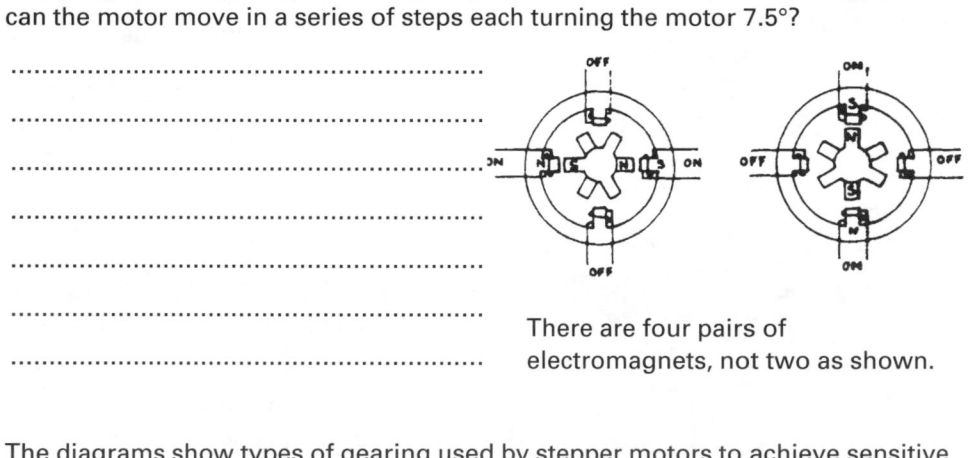

....

....

....

....

....

....

There are four pairs of electromagnets, not two as shown.

The diagrams show types of gearing used by stepper motors to achieve sensitive linear or rotating motion. Identify the uses of the stepper motors shown.

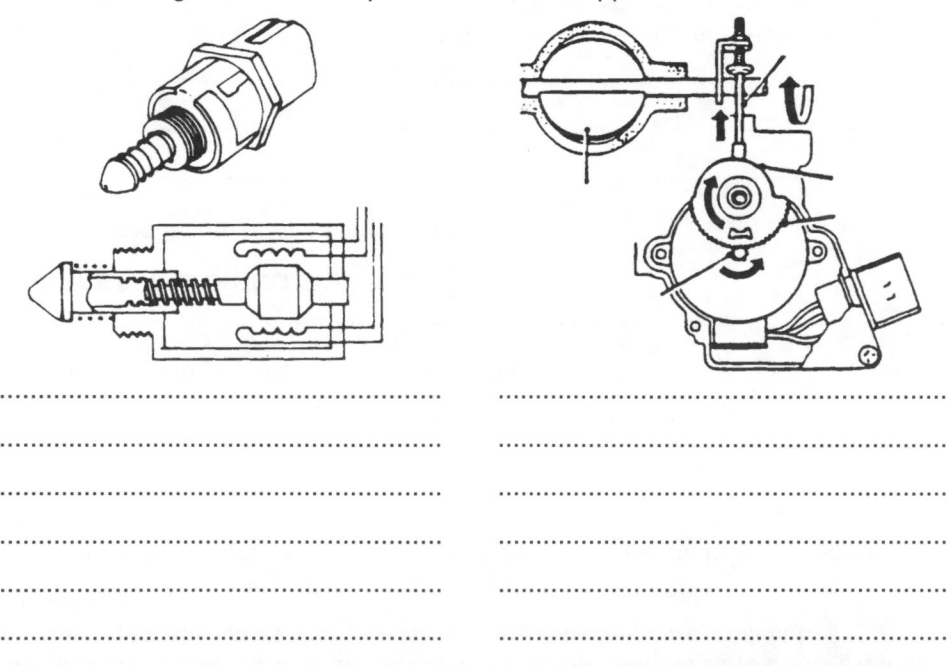

....

....

....

....

....

....

DC MOTORS

The permanent-magnet DC motor is the most common type of actuator used on cars to produce rotating movement; or by suitable gearing, reciprocating or linear motion. It is known as an Iron-cored Motor.

SERVO MOTORS

How do these motors differ from the conventional DC type?

....

....

....

....

Describe the motor shown.

fixed permanent magnet — rotating armature and windings — commutator and brushes

....

....

....

FUEL FLOW METER

The diagram shows the principle of operation of one type of fuel flow meter. Describe how it measures the flow rate of fuel.

....

....

....

....

....

TESTING EQUIPMENT

When testing the discrete and integrated components shown in this chapter, it is necessary to determine that there is a correct input and output voltage and that the resistance values are correct.

Below is shown typical equipment. Identify the instruments and state their function.

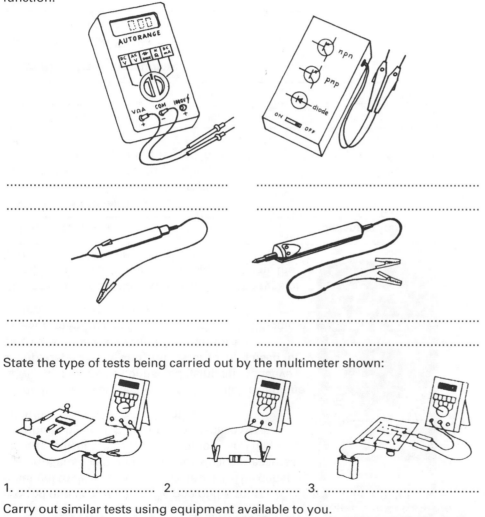

....

....

....

State the type of tests being carried out by the multimeter shown:

1. ....   2. ....   3. ....

Carry out similar tests using equipment available to you.

DIAGNOSING VEHICLE SENSOR AND CIRCUIT FAULTS

The microprocessor which receives signals from the sensors cannot be tested; therefore all other items in the particular circuit must be tested for correct operation before considering the microprocessor to be faulty.

This means a diagnostic procedure consisting of a series of simple checks on sensors, warning-lights/displays, wiring connectors and wiring assemblies.

There are FOUR guiding principles to any diagnosis of vehicle electronic system faults:

1. ....

2. ....
....
....

3. ....
....
....

4. ....

Examine a vehicle's data specification and complete the table below

VEHICLE MODEL	
Component	Expected value
Coolant temperature sensor	
Air temperature sensor	
Crankangle sensor	
Fuel pressure	
Injectors	
MAP sensor	
Lambda sensor	

AUTOMOTIVE MULTIMETER

A good quality multimeter is one of the most essential tools in a technician's tool kit. The general multimeter shown on the previous page measures voltage, current, ohms, checks diodes and has a continuity bleeper, but for motor vehicle work additional functions are essential.

Some functions are self-explanatory such as **Engine RPM**, which may be connected using an inductive clamp, **Capacitance and Transistor testing** used for electronic repair work, and **Temperature measurement** to check cooling and air conditioning systems, and exhaust catalysts.

Other functions are named below. State the type of sensor or item each function checks and name the parts indicated on the multimeter.

©GUNSON

Multimeter function	Sensor or item
Duty Cycle / Dwell This is a measurement of the time a component is switched on expressed as a % (Dwell is a duty cycle expressed in degrees)	
Pulse width	
Low AC Voltage measurement	
Bargraph display Shows movement similar to an analogue display	
Data Hold Function Will hold minimum and maximum values to be recalled later.	
RS232 Link	

SYSTEM PROTECTION AND PRECAUTIONS

List under the following headings the general rules/precautions to be observed to ensure that electronic components have a long life and are protected during testing and replacing:

1. Excessive use of heat ....

....

....

2. Connecting electrical supply ....

....

....

....

3. Damage during use ....

....

....

....

....

....

....

List the principal test equipment used/available during the practical work on this element:

....

....

....

What precautions should be taken when using such equipment?

....

....

....

SERIES CIRCUITS – EFFECTS OF RESISTORS

State OHM'S LAW as a formula:

....

....

....

It is possible to represent a circuit by either a 'conventional' or 'electronic' diagram – as can be seen opposite.

State the laws of series circuits, and for the circuit shown calculate: resistance, current and voltage.

RESISTANCE ....

....

....

CURRENT ....

....

....

....

....

VOLTAGE ....

....

....

....

....

....

State the SI symbols for:

Resistance ....

Voltage ....

Current ....

CONVENTIONAL CIRCUIT   ELECTRONIC CIRCUIT

1. Two resistors of 1.75 Ω and 4.25 Ω are connected in series.

   What voltage would be required to cause a current of 2.5 A to flow in the circuit?

....

....

....

....

....

....

2. Four resistors of 6,8,10 and 12 Ω are connected in series to a 12 V circuit.
   Calculate the total resistance of the circuit and the current flowing in each resistor.

....

....

....

....

....

....

3. Four resistors of equal value are placed in series and connected to a 110 V supply. A current of 5 A then flows.

   Calculate the value of each resistor and the voltage across each resistor.

....

....

....

....

....

....

4. Three resistors are wired in series and when connected to a 12 V supply a current of 6 A flows in the circuit.

   If two of the resistors have values of 0.5 and 0.8 Ω, calculate the value of the third resistor.

....

....

....

....

....

....

PARALLEL CIRCUITS – EFFECTS OF RESISTORS

State the basic laws of parallel circuits; and for the circuit shown calculate: resistance, current and voltage.

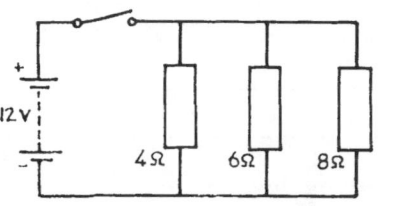

CONVENTIONAL CIRCUIT   ELECTRONIC CIRCUIT

RESISTANCE ....

....

....

....

....

....

....

CURRENT ....

....

....

....

....

....

....

VOLTAGE ....

....

PROBLEMS

1. Two resistors of 20 and 5 Ω are connected in a 12 V parallel circuit.

   Calculate the total resistance of the circuit.

   ....

   ....

   ....

   ....

   ....

   ....

   ....

3. Three resistors of 3, 5 and 6 Ω are connected to a 12 V battery.

   Calculate the total circuit resistance.

   ....

   ....

   ....

   ....

   ....

   ....

2. Three conductors are placed in a parallel circuit, their resistances being 2, 3 and 4 Ω. Calculate the total resistance of the circuit when connected to a 12 V system.

   ....

   ....

   ....

   ....

   ....

   ....

4. Two resistors of 4 and 8 Ω are connected in parallel. They are then connected in series to a 3.33 Ω resistor.

   Calculate the total resistance of the circuit.

   ....

   ....

   ....

   ....

   ....

   ....

ELECTRICAL TESTING PROBLEMS

State with a letter **A** or **V**, to indicate on the drawings which meters are Ammeters or Voltmeters.

1. State the reading that would be obtained on each meter if the coil has a resistance of 4 ohms.

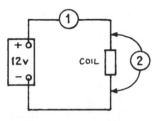

What would happen if the two meters were changed over in the circuit?

Meter 1: ....

Meter 2: ....

2. Calculate the readings that would be shown on meters 1 and 2.

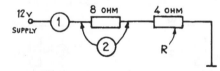

State how the voltmeter could be used to confirm an open circuit in resistor R.

....

....

....

3. Calculate the readings that would be obtained on meters 1, 2 and 3.

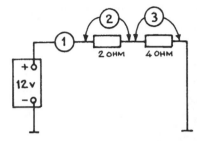

Calculate the power consumed in the above circuit.

4. Calculate the readings that would be obtained on meters 1, 2 and 3.

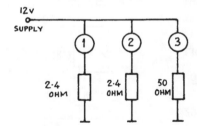

Calculate the equivalent resistance of the above circuit.

5. (a) Calculate the current flowing in resistor A.

   (b) Calculate the current that would flow in resistor C when the switch is (i) open, (ii) closed.

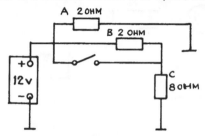

6. State the readings that would be obtained on each meter.

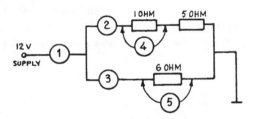

SERIES CIRCUITS IN ELECTRONIC SYSTEMS

POTENTIAL DIVIDER

This divides potential (voltage) in a series circuit and so provides a means of controlling outputs as they vary.

The diagrams show three similar circuits having different resistor values. These values are much greater than those in the examples on the previous pages. Their large variations provide a more sensitive control.

12V

1kΩ R_1
6Vout
1kΩ R_2 (v)

0V

The resistance values are the same. This means the voltage across each resistor will be the same.

Calculate the above values using the formula:

For example:

$$V\text{out} = \frac{1}{1+1} \times 12$$

$$\frac{1}{2} \times 12$$

$$= 6V$$

How can this voltage output provide control?

....

....

....

....

12V

1kΩ R_1

5kΩ R_2 (v)

0V

The resistance value R_2 is greater than R_1. This means the voltage across R_2 will be .... than the voltage across R_1.

$$\text{Voltage out} = \frac{R_2}{R_1 + R_2} \times \text{Supply voltage}$$

12V

11kΩ R_1

1kΩ R_2 (v)

0V

The resistance value R_2 is less than R_1. This means the voltage across R_2 will be .... than the voltage across R_1.

TEMPERATURE SENSOR

An engine-cooling system uses a thermistor to indicate temperature change.

Calculate the 'voltage out' values when the engine is cold and then hot. The thermistor resistance is 160 k Ω cold and 2.5 k Ω hot.

Note: a voltage regulator has reduced the battery voltage to an input of 9 V.

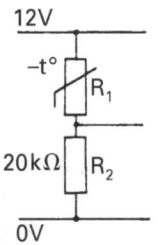

12V

−t° R_1

20kΩ R_2

0V

LIGHT SENSOR

A light sensor could be used for automatically switching on vehicle lights as it becomes dark. The LDR (light dependent resistor) changes its resistance value as the level of light falling on its 'window' changes.

Calculate the 'voltage out' values in daylight and darkness when the LDR values are 1 k Ω in daylight and 10 M Ω in darkness.

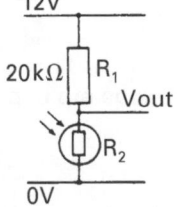

12V

20kΩ R_1
Vout

R_2

0V

Complete the diagram to show a control circuit that would operate a light.

Note: After passing control, the voltage is further reduced by the 1 k Ω resistor as it passes to the transistor.

Approximately what voltage is required to trigger a transistor base? ....

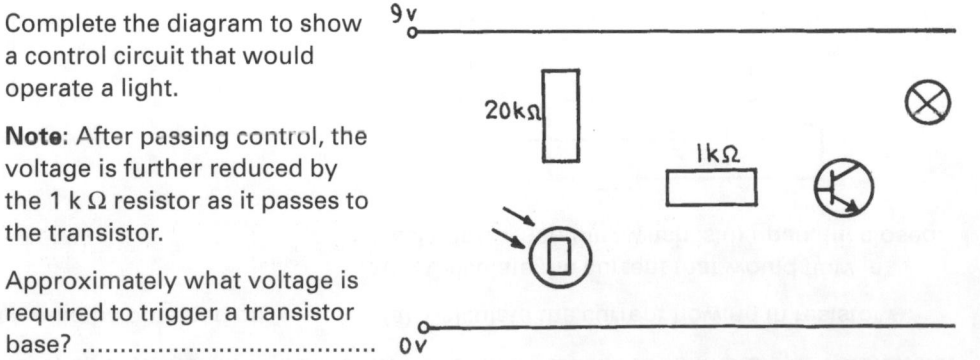

9v

20kΩ

1kΩ

0v

Chapter 12

Engine Management & Petrol Fuel Systems

Petrol injection systems – Bosch   252
Engine management systems   253
Fuel system layout   253
Component location for electronic fuel injection systems   254
Electronic control unit (ECU)   255
Control of spark timing   255
Engine process maps   256
Open- and closed-loop engine control systems   257
Programmed ignition systems   258
Saab direct ignition (DI)   259
Electronically controlled carburettors   260
Stepper motor   261
Diagnostics   262
Fuel injection systems   263
Bosch K Jetronic system layout   263

Bosch KE Jetronic system layout   267
Bosch L Jetronic system layout   268
Bosch LE3 Jetronic system layout   270
Bosch Motronic system layout   271
Multi-point fuel injection system (Rover)   272
Single-point fuel injection system (Rover)   273
Bosch Mono-Jetronic system layout   274
Lambda sensor   275
Exhaust gas recirculation (EGR)   275
Engine running problems   275
Diagnostics   276
Engine self diagnosis   277
Diagnosing difficult faults   278
System protection during use   279

PETROL INJECTION SYSTEMS – BOSCH

Complete the sketches below by naming the main parts and describe the basic operation of each.

K/KE-JETRONIC

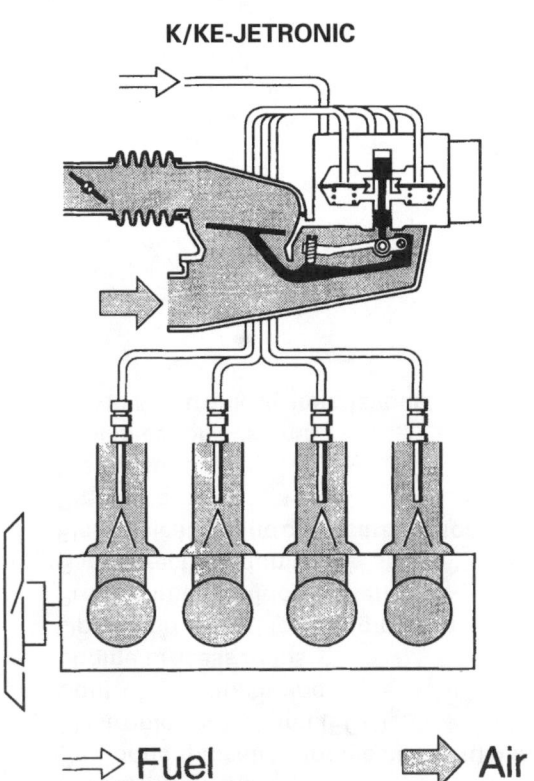

⟹ Fuel   ⇨ Air

K/KE-Jetronic is a mechanical electronic system. K stands for kontinuerlich, the German word for continuous. The fuel is continuously injected when the engine is running.

....

....

....

....

....

SINGLE POINT INJECTION

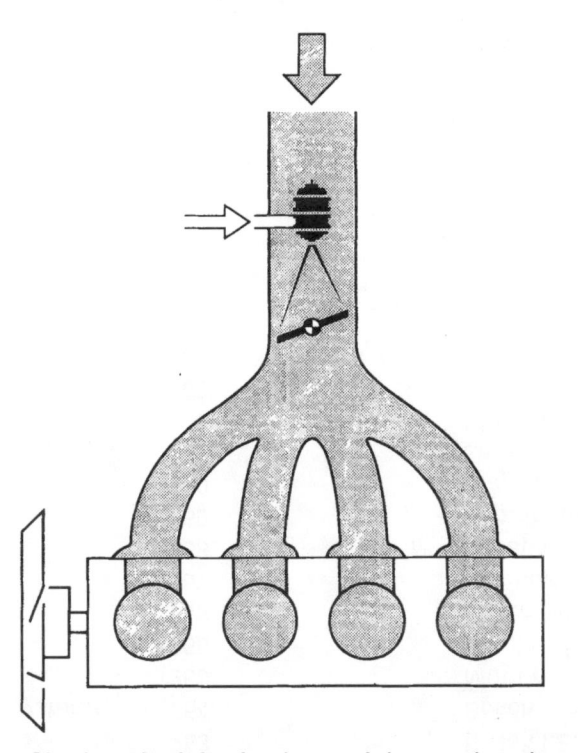

Single point injection has an injector placed at about the normal carburettor position and is very suitable for vehicles below about 1.8 litres capacity.

....

....

....

....

MULTI-POINT INJECTION

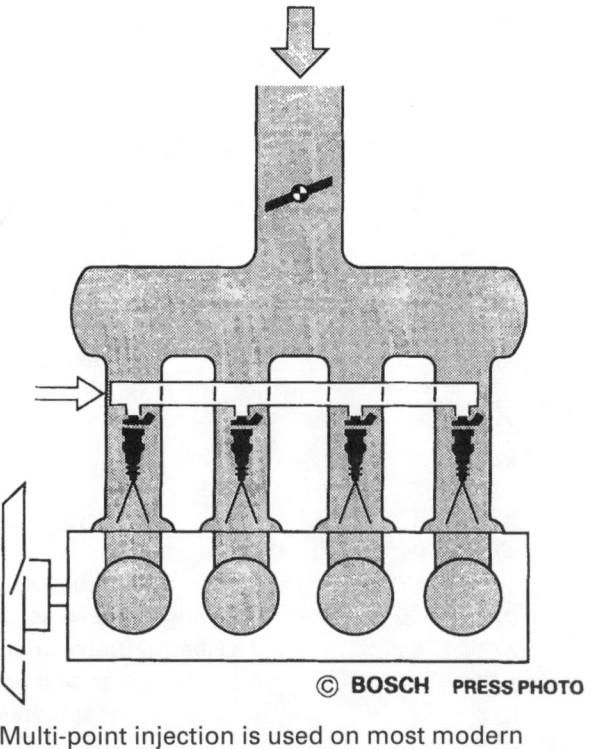

© BOSCH PRESS PHOTO

Multi-point injection is used on most modern systems; each inlet port has an injector fitted near to the inlet valve(s). All injectors have an injection pulse: (1) at the same instant every engine rev – L/LE/LH Jetronic or (2) sequentially – some Motronic systems.

....

....

....

ENGINE MANAGEMENT SYSTEMS

Engine management is a simplified term used to describe the electronic control of a running engine during all its varying speeds and loadings.

A spark-ignition engine may use a carburettor or fuel injection to distribute fuel while the spark will be controlled by some form of electronic ignition.

State some types of engine management control:

1. ....

2. ....

....

....

....

3. ....

4. ....

....

....

The advantages gained using engine management systems compared with the basic ignition and carburettor systems are:

(a) ....

(b) ....

(c) ....

(d) ....

These are due to:

(i) ....

(ii) ....

(iii) ....

(iv) ....

....

FUEL SYSTEM LAYOUT

A simplified block diagram layout of a modern full flow pressure recirculating fuel injection system is shown below. The full system layout which includes the venting system can be seen on the vehicle illustrated on the next page. The system is pressure regulated to approximately 270 kPa (2.7 bar).

Name the parts indicated.   © **ROVER**

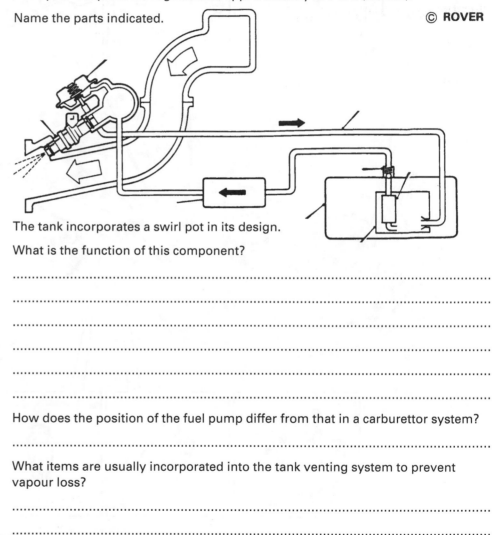

The tank incorporates a swirl pot in its design.

What is the function of this component?

....

....

....

....

....

How does the position of the fuel pump differ from that in a carburettor system?

....

What items are usually incorporated into the tank venting system to prevent vapour loss?

....

COMPONENT LOCATION FOR ELECTRONIC FUEL INJECTION SYSTEMS

The fuel injection system shown is that used on Austin Rover O-series engines.

The drawing shows the relative positions of the components.

Recognise and name the numbered parts.

1. ....
2. ....
3. ....
4. ....
5. ....
6. ....
7. ....
8. ....
9. ....
10. ....
11. ....
12. ....
13. ....
14. ....
15. ....
16. ....
17. ....

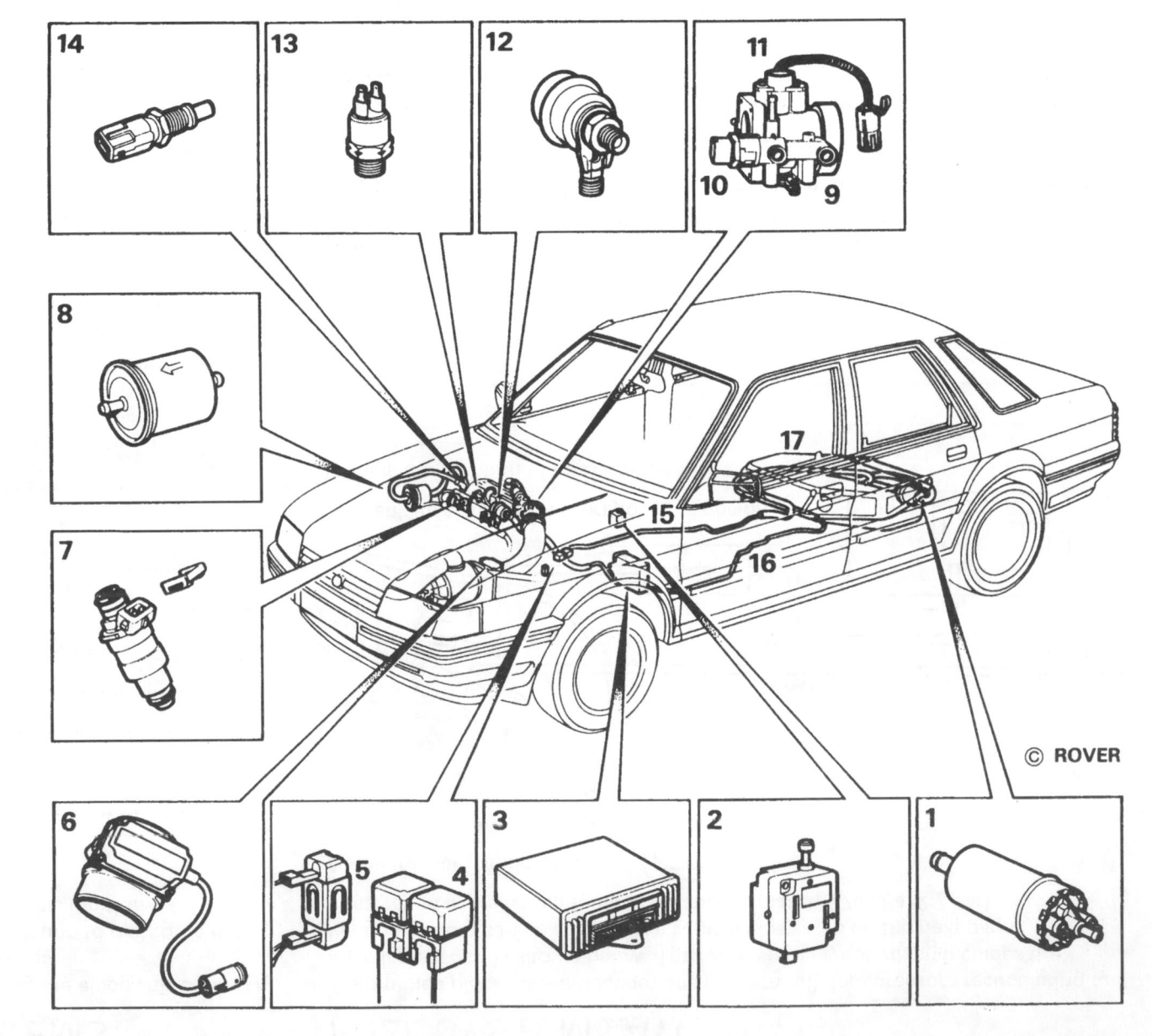

© ROVER

ELECTRONIC CONTROL UNIT (ECU)

The electronic control unit for an engine management system is a small computer and, as all computers, is made up of integrated circuits, diodes, resistors and capacitors.

What is the ECU's basic function?

....

....

....

Complete the diagram by naming the typical items which send information to the ECU.

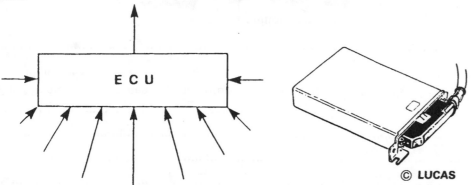

© LUCAS

The actual fuel injection time is known as a pulse width. How will this alter under the various running conditions?

....

....

....

Current ECUs include a 'limp home' mode. What does this mean?

....

....

....

CONTROL OF SPARK TIMING

In order to ensure that the pressure build-up in the combustion chamber occurs just as the piston passes tdc the spark must be timed to occur approximately 5° to 10° before tdc when the engine is turning at idling speeds. The three basic factors affecting variation of the spark firing from this position are:

SPEED

....

....

....

....

LOAD

....

....

....

....

....

AIR–FUEL RATIO

....

....

State the effects of incorrect ignition timing for a set engine speed.

1. Over-advanced ....

....

....

2. Retarded ....

....

....

ENGINE PROCESS MAPS

SPARK ADVANCE MAP

The three-dimensional map shown below shows the correct ignition timing for an engine when it was tested over the whole range of speeds and loads, the ignition timing being varied until the best results were obtained. It indicates optimum timing settings that should be used throughout the engine's speed range.

How are such ignition timing values used?

....
....
....
....
....
....

What items does this memory map (Bosch) replace in the conventional mechanical distributor?

....

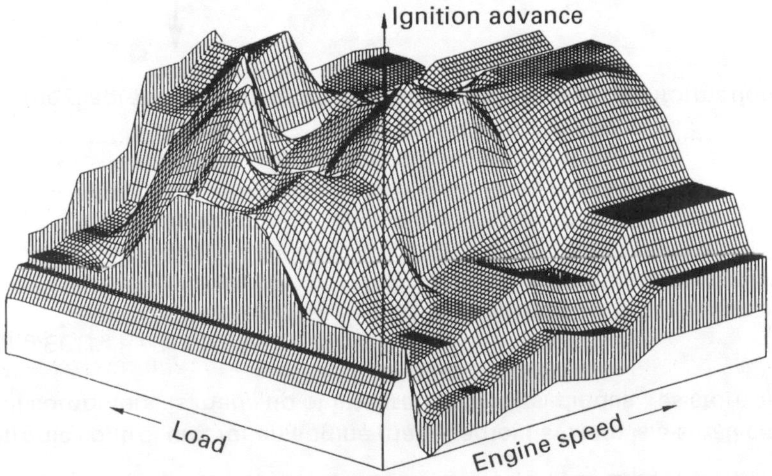

Ignition advance

Load

Engine speed

FUEL MIXTURE MAP

The three-dimensional map (Bosch) below shows how the air–fuel ratio must change with speed and load to give a constant chemically correct value.

The map indicates (1000 x 1000) one million 'look-up' points that the ECU can detect and reset to.

This map will only work properly for an engine when the speed is reasonably constant and the engine is at normal running temperature.
What other sensors must be provided to allow the engine to respond to the driver via the ECU?

....
....
....
....
....
....
....
....

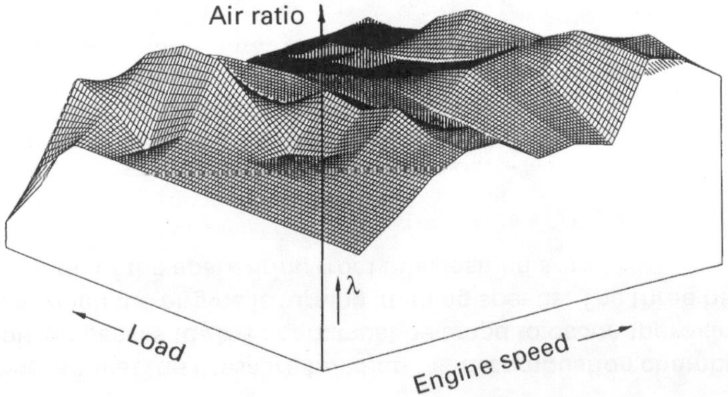

Air ratio

λ

Load

Engine speed

© **BOSCH**

OPEN-LOOP ENGINE CONTROL SYSTEM

An engine receives control signals to indicate, from the ignition system, at what point to fire the mixture; and from the fuel system, how much fuel to supply. The open-loop system monitors these signals to run the engine. However it does not receive self-correcting feedback via actual engine output which could indicate the need for constant minor adjustment.

Complete the blocks to show an ECU controlling an open-loop spark-ignition petrol-engine system.

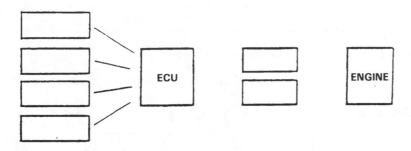

With the engine running, the ECU receives signals from sensors. State where these sensors are positioned to indicate:

Engine speed ....

Engine load ....

....

Engine temperature ....

Ambient temperature ....

....

Using these signals the ECU calculates the amount of fuel required and the correct spark plug firing point.
What is the disadvantage of the open-loop system?

....

....

....

CLOSED-LOOP ENGINE CONTROL SYSTEM

This is similar to the open-loop system, but the output of the engine in terms of exhaust emissions and engine combustion knock is measured and fed back to the ECU to give an improved performance. The output sensors only signal if combustion knock (pinking) occurs or if the exhaust emissions indicate that the air–fuel mixture is too rich. This allows the process maps to be designed to the limit of engine design and economy.

Complete the blocks to show an ECU controlling a closed-loop spark-ignition petrol-engine system and describe the action of the output sensors.

KNOCK DETECTION
The knock detector is mounted in the upper part of the cylinder block.

....

....

....

....

....

EXHAUST EMISSION DETECTION
The lambda sensor is positioned in the exhaust manifold or before the catalytic burner.

....

....

....

....

PROGRAMMED IGNITION SYSTEMS

The Lucas digital electronic ignition system, as used by some Rover Group models, is shown below. The high-tension circuit is conventional; the rotor is positioned in a vibration-absorbing bush in the end of the overhead camshaft and the distributor cap is mounted on to it on the cylinder head.

Name the major parts numbered 1 to 7.

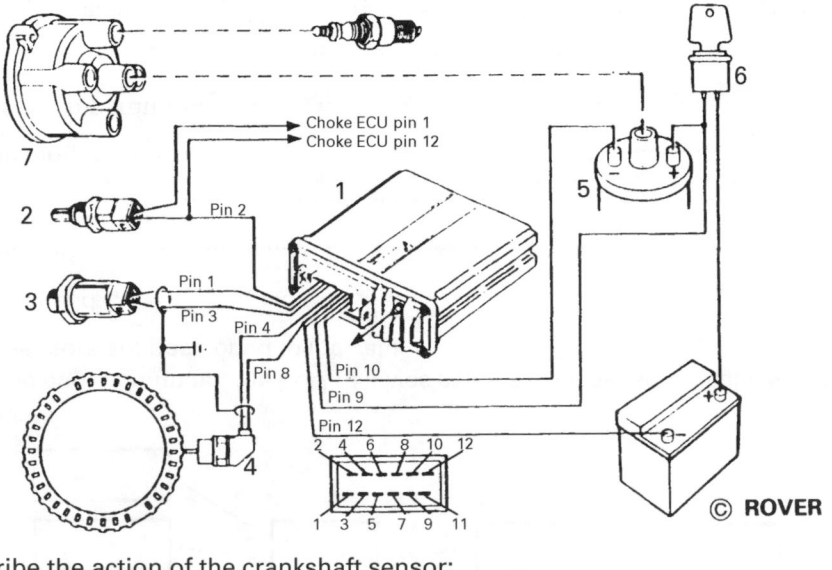

© ROVER

Describe the action of the crankshaft sensor:

....
....
....
....
....
....
....
....

DISTRIBUTORLESS IGNITION SYSTEM (DIS) used by Ford

The flywheel has (36–1) 35 teeth. The missing tooth space passes the variable reluctance sensor (VRS) at a point 90° btdc of No. 1 cylinder. This positioning, and the speed of the pulse generated as the flywheel turns, give information for spark plug firing and engine speed.

Name the parts indicated.

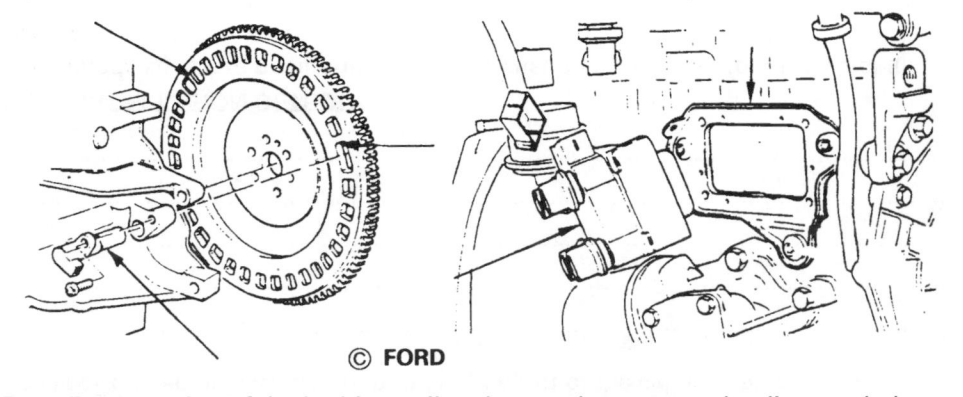

© FORD

Describe the action of the ignition coil and name the parts on the diagram below:

....
....
....
....
....
....
....

© FORD

SAAB DIRECT IGNITION (DI)

The SAAB DI is a capacitive ignition system, it requires no distributor, no moving parts and no ignition (plug) leads. Each sparking plug is equipped with its own, very small, ignition coil. The ECU contains pre-set data which controls, as well as the ignition, the fuel injection and the turbocharger operation.

Name the parts indicated on the ignition layouts shown on this page.

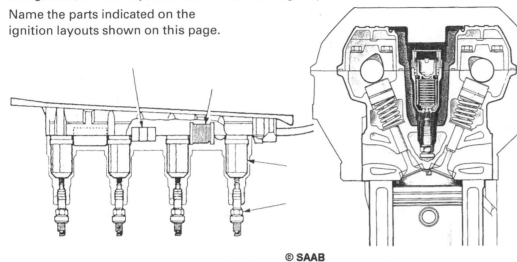

© SAAB

Describe how the voltage build up to the sparking plugs is in two stages.

....

....

....

....

....

....

The voltage of 40000v, at high speeds allows not one but a blast of sparks across the electrodes as each plug fires. State the advantages of this high voltage, clean-burning effect.

....

....

....

State the advantages that the ignition cartridge cover provides compared to a system using ignition leads.

....

....

....

© SAAB

On this DI system how is the engine piston position and speed determined?

....

....

....

....

....

To determine which cylinder is on the firing stroke a continuous low voltage (80V) is supplied to all the sparking plugs, and since the electrical resistance of a gas decreases as it is heated, the ECU can tell which cylinder is near the top of its compression stroke because the current at that cylinder starts to ionise across the gap. This current flow also acts as a knock sensor. Any unusual occurrence in the combustion process will upset the ions flow and the ECU will sense and control the cylinder knock. The system can detect the changes during the injection period and make corrections before the injection pulse stops.

What is the function of the pressure sensor in the inlet manifold?

....

ELECTRONICALLY CONTROLLED CARBURETTORS

The mechanically controlled carburettor has become so complicated in its use with high-speed engines that it has difficulty in coping with modern emission requirements. If electronic control is used to aid the carburettor's functions, a vast improvement to the vehicle's running can be made and it will stay in tune for longer periods. The types of electronic carburettor in current use are known as hybrids since they are basically a conventional mechanical carburettor with controls that are operated via an ECU.

What systems on the carburettor may be controlled by the ECU?

....

....

....

....

Opposite is shown an SU HIF carburettor. Describe what is controlled from the ECU on this unit.

....

....

....

....

....

....

....

SU CARBURETTOR ELECTRONIC MIXTURE CONTROL CIRCUIT (ROVER 216)

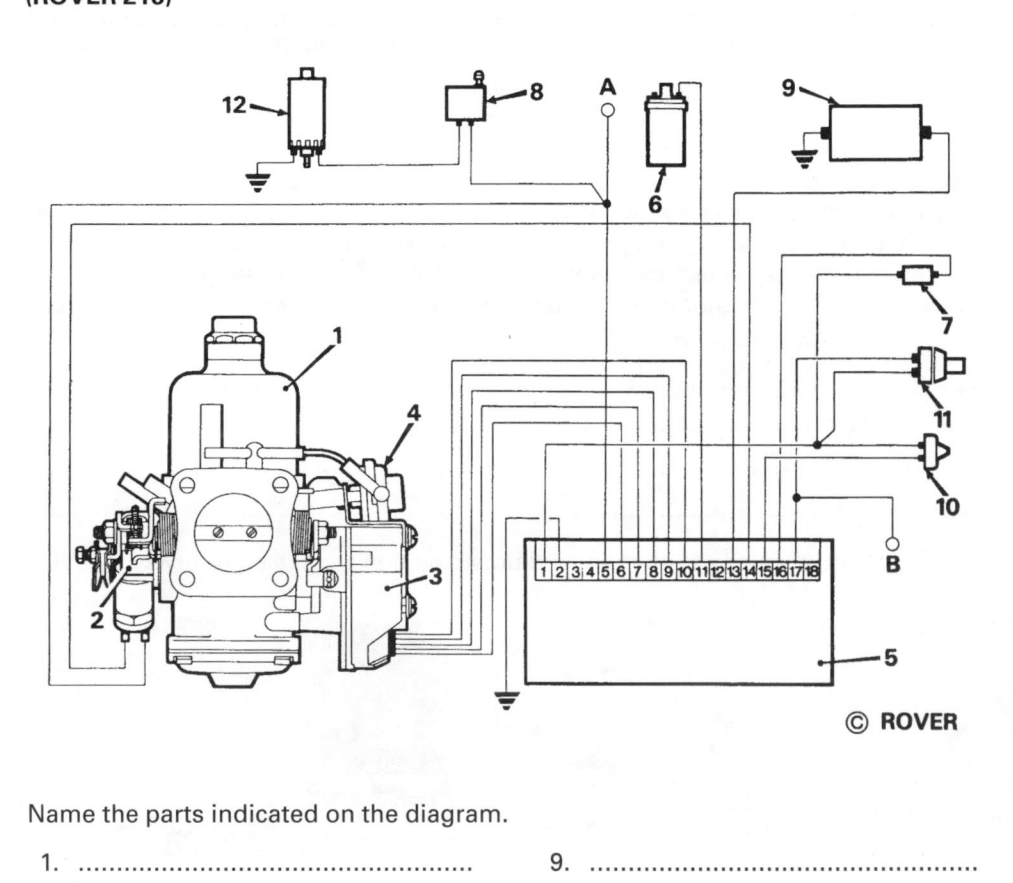

© ROVER

Name the parts indicated on the diagram.

1.	9. ....
2.	10. ....
3.	11. ....
4.	12. ....
5.	
6.	A ....
7.	B ....
8.	

STEPPER MOTOR (See also page 244)

This type of motor turns a specified amount (in steps) and then holds its position. It does not spin like a conventional electric motor.

The type of stepper motor used on the SU carburettor is of the permanent magnet type and has basic step-movement angles of 7.5°. Other designs may have steps down to 1.5°.

Describe the action of the SU carburettor stepper motor to control the rotary choke.

....
....
....
....
....
....
....
....

What is the maximum rotation that can be made by the disc valve? ....

What is the advantage of using a permanent magnet type stepper motor?

....
....

Name the items on the simplified choke layout shown, and indicate the mixture flow.

What is an auxiliary carburettor?

....
....

fuel from float chamber

stepper motor

air

Drawings show stepper motor fitted to the Weber 2 V carburettor fitted on some Ford Granada models.

Name items idicated.   Identify stepper motor on carb.

WEBER 2 V CARBURETTOR

STEPPER MOTOR HOUSING

Describe the basic action of this stepper motor.

....
....
....
....
....
....
....
....
....
....
....

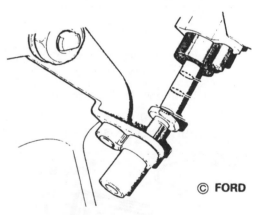

© **FORD**

Show stepper motor plunger positions for:

(A) vent manifold/start
(B) idling
(C) anti-dieselling/anti-run-on.

The plunger action can also be controlled by a vacuum-operated diaphragm, the vacuum being switched on and off by the ECU.

ENGINE PERFORMANCE – DIAGNOSTIC TESTING FOR ENGINE MANAGEMENT AND PETROL FUEL SYSTEMS

List the types of tests that may be carried out to determine the proper running of an engine:

1. ....

....

2. ....

....

3. ....

4. ....

....

5. ....

6. ....

....

Note: Details based on the tests/checks 1–4 have already been covered in other sections of the book. You will, however, in Workshop classes still be expected to describe such tests when completing this element of work.

FAST CHECK TESTING

Using this type of diagnostic equipment, a technician can quickly and accurately diagnose electrical faults. Give details.

....

....

....

....

....

....

....

IGNITION   © ROVER

....

....

....

....

....

....

....

....

....

BREAKOUT BOX TESTS

As well as using quick testers, Ford use a breakout box in conjunction with a multimeter to test the various systems.

....

....

....

....

Name the numbered parts:

1. ....

2. ....

3. ....

© ROVER

ELECTRONIC CARBURETTOR
FAST CHECK

© FORD

FUEL INJECTION SYSTEMS

FUEL DELIVERY – MECHANICAL

There are two types of mechanical fuel injection systems: those that require a drive from the engine and those that do not. The engine-driven systems use a fuel injection pump and governor very similar to the fuel injection pumps used on diesel engines. These however have become less common for petrol engine use.

The other mechanical system is one which injects fuel continuously and requires no mechanical drive. It is considered a mechanical system because basic fuel metering is controlled by the mechanical relationship between the air-flow sensor (10) and the control plunger in the fuel distributor (9a).

BOSCH K JETRONIC SYSTEM LAYOUT

CONTINUOUS INJECTION SYSTEM (CIS)

The air-flow measurement system and the fuel delivery system interact in the mixture control unit and provide a correct air–fuel mixture to be supplied via the intake tubes.

Complete the block diagram below to show:

1. Air flow, 2. Fuel supply, 3. Fuel induction

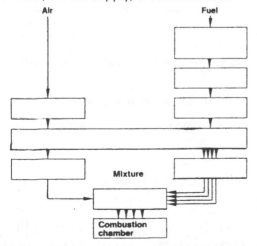

Examine the Bosh K Jetronic schematic diagram shown below and name the numbered parts.

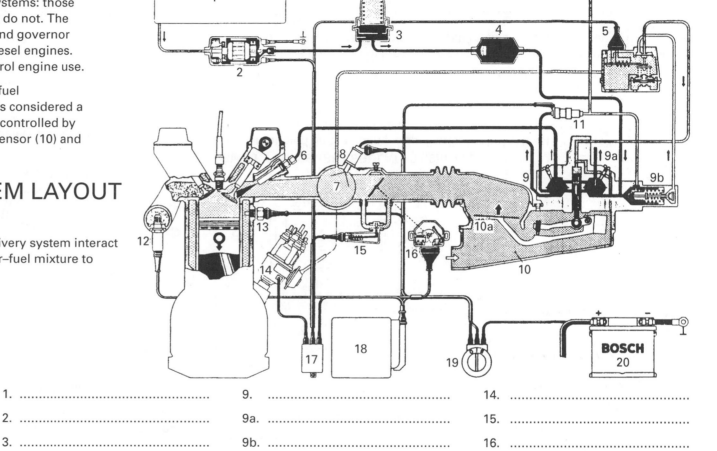

1.	9.	14. ....
2.	9a.	15. ....
3.	9b.	16. ....
4.	10.	17. ....
5.	10a.	18. ....
6.	11.	19. ....
7.	12.	20. ....
8.	13.	

Functions and Operation of Fuel Injection Parts

FUEL SUPPLY

An electrically driven fuel pump is gravity-fed fuel from the tank. The fuel is forced by the pump through a pressure accumulator and the fine filter to the fuel distributor.

ELECTRIC FUEL PUMP ((2), page 263)

The fuel pump is a (Bosch) roller cell type which is driven by a permanent-magnet electric motor.

Name the parts indicated on the drawings and describe the pump's operation:

....

....

....

© BOSCH

What feature makes this pump different from other electric fuel pumps?

....

....

....

What safety feature is employed in the pump's system?

....

....

....

FUEL ACCUMULATOR ((3), page 263)

State reasons why a fuel pressure accumulator is required in this system.

....

....

....

....

....

....

....

....

....

Name the main parts.

ACCUMULATOR SHOWN FULLY
PRESSURISED   © BOSCH

FUEL FILTER ((4), page 263)

The paper filter element is larger than a carburettor's element; it has a strainer nylon screen fitted at one end to catch any loose particles. When positioning:

....

....

....

....

PRIMARY PRESSURE REGULATOR ((9b), page 263)

This is mounted on the main pump distributor body. It acts in a similar manner to an engine oil pressure relief valve, for example:

....

....

In a continuous flow system the control of fuel pressure is more important than in any of the other types of fuel injection system, this is because:

....

....

....

MIXTURE CONTROL UNIT ((9–10), page 263)

The mixture control unit is where the air-flow measurement system and the fuel delivery system interact. The engine's air-flow intake is measured and a metered proportion of fuel is then supplied to the injectors. The mixture control unit is therefore a combination of two separate components which are the:

.... and ....

Name the parts indicated.

UPDRAFT AIR-FLOW
SENSOR

© BOSCH

DIFFERENTIAL
PRESSURE VALVE

Describe the fuel supply operation:

....

....

....

....

....

....

....

....

....

....

IDLE SPEED ADJUSTMENT
Idle speed is adjusted by controlling the small amount of air flow that is allowed to by-pass the throttle. Idle speed adjustments do not affect the mixture adjustment because the air flow has already been measured. Adjust by:

....

MIXTURE CO ADJUSTMENT
Why does turning the idle mixture screw alter the CO setting?

....

....

....

How is good acceleration response achieved?

....

....

ENGINE WARM-UP

Air–fuel enrichment is achieved in three ways or with the aid of three devices:

1. WARM-UP REGULATOR ((5), page 263)

This is an electrically controlled pressure valve which is open when the engine is switched on and progressively closes during warm-up. What does this closing action achieve?

....

....

....

....

....

2. AUXILIARY AIR DEVICE ((15), page 263)

This is a perforated plate operated by a bi-metal spring. It is heated electrically, which allows its opening time to be restricted depending on engine type. The device is positioned on the engine so that it keeps warm; this prevents it from opening during warm engine restarts.

Name the main parts of the auxiliary air device shown below and describe when it allows extra air flow.

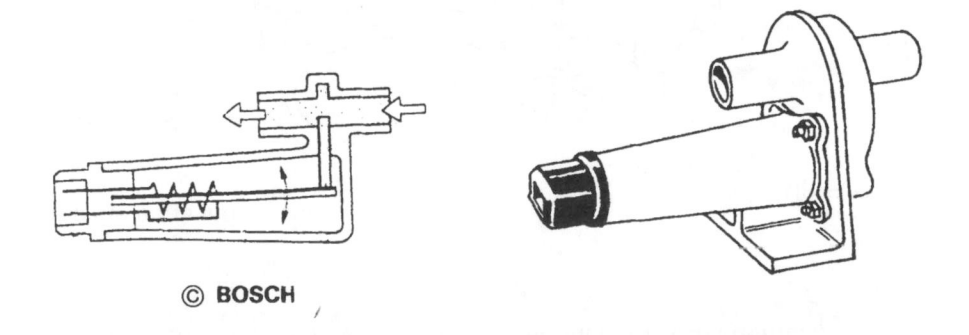

© BOSCH

....

....

....

3. COLD START VALVE ((8), page 263) and THERMO TIME SWITCH ((13), page 263)

The cold start injector is energised only if the engine is cold. It cannot be activated, to cause possible flooding, when the engine is warm.

When the ignition key is turned to start, current supplied from the starter solenoid during cranking is passed through the cold start valve, it is activated, and the current flow is grounded through the thermo time switch positioned in the engine block. Briefly describe how they work.

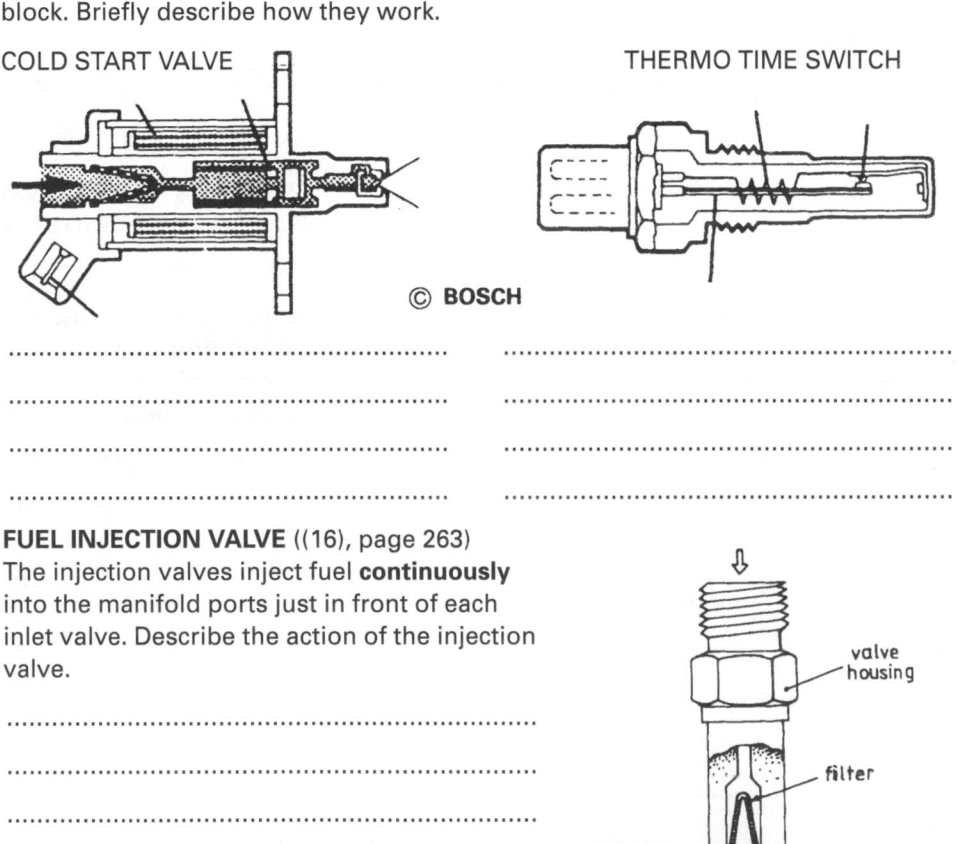

COLD START VALVE   THERMO TIME SWITCH

© BOSCH

....

....

....

....

FUEL INJECTION VALVE ((16), page 263)

The injection valves inject fuel **continuously** into the manifold ports just in front of each inlet valve. Describe the action of the injection valve.

....

....

....

....

....

....

valve housing

filter

valve needle

valve seat

valve closed   valve open

BOSCH KE JETRONIC SYSTEM LAYOUT

CONTINUOUS ELECTRONIC SYSTEM

The mechanical parts of this system are similar to the K-Jetronic system, but with the added use of Electronic Control to increase flexibility and give additional control options.

ELECTRO-HYDRAULIC PRESSURE ACTUATOR

The main additional feature to the system is the use of the electro-hydraulic pressure actuator fitted to the fuel distributor. The actuator receives signals from the electronic control unit (ECU) to modify the fuel supply when required.

How does the pressure actuator control fuel pressure variations?

....

....

....

....

What control options are gained by using this pressure actuator?

1. ....

2. ....

3. ....

FUEL DISTRIBUTOR
WITH ELECTRO-HYDRAULIC
PRESSURE ACTUATOR

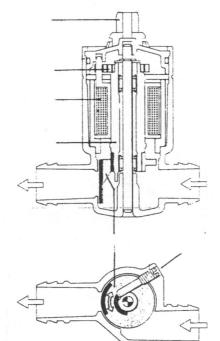

Identify the various parts.

© BOSCH

THROTTLE VALVE SWITCH

This is connected to the throttle valve body and operated by the throttle valve shaft.

State its two important features:

....

....

....

....

....

....

....

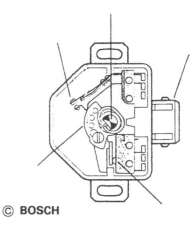

© BOSCH

ROTARY IDLE ACTUATOR

This actuator when fitted replaces the electrically heated air device (15), page 263). What is its basic function?

....

....

....

....

....

....

....

The signals are corrected on a closed loop which involves reading variable values from various sensors, e.g. engine speed, temperature, throttle position.

When would the idle speed require stablising?

....

....

....

....

Fuel delivery–electronic pulse

The basic difference between the mechanical system already described and the electronic system is that in the electronic system the injection valves are solenoid operated and controlled by the ECU. The amount of fuel injected depends on the length of time the injectors are allowed to stay open.

BOSCH L JETRONIC SYSTEM LAYOUT

This is an electronic system. The fuel pump, filter, solenoid and many of the sensors are similar to the mechanical system already described.
The fuel is pulsed into the inlet manifold by the injection valves once every engine rev (i.e. twice during each four stroke cycle).

Complete the much simplified block diagram below to show:

1. Air flow   2. Fuel supply   3. Electronic control

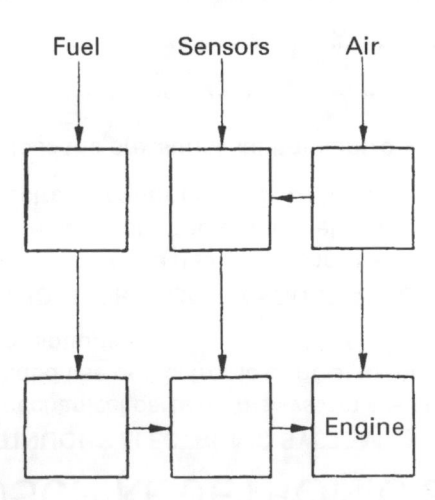

Examine the Bosch L Jetronic schematic diagram shown above and name the numbered parts.

1. ....
2. ....
3. ....
4. ....
5. ....

6. ....
7. ....
8. ....
9. ....
10. ....

11. ....
12. ....
13. ....
14. ....
15. ....

Functions and Operation of Fuel Injection Parts

FUEL SUPPLY

From the fuel filter, the fuel is supplied under pressure into the fuel rail; any excess is returned to the tank through the pressure regulator which maintains the system at a pressure of 2.5–3 bar (250–300 kPa) depending on type.

How are fuel pulsations absorbed?

....

....

....

© BOSCH

Name all the parts indicated and describe their operation.

FUEL INJECTION VALVE

The valves are solenoid operated, and are opened and closed by electric impulses from the ECU. (The fuel supply is not continuous.)

....

....

....

....

....

....

....

PRESSURE REGULATOR

To ensure that the injector delivers a precise amount of fuel for each millisecond that it is open, the pressure regulator must maintain a constant relative pressure at the injector tip of 2.5 bar. This means that the fuel pressure in the fuel line must vary by the same amount as the manifold pressure varies during all the varying engine running conditions.

Describe the initial operation of the pressure regulator shown:

....

....

....

....

....

....

....

....

....

© BOSCH

Under running conditions how does the pressure regulator maintain a constant relative pressure at the injector tip when the manifold pressure is constantly varying owing to the changing engine load requirements?

....

....

....

....

© BOSCH

Why is maintaining a constant pressure so important?

....

....

AIR-FLOW MEASUREMENT

A simplified
diagram of the
air-flow intake
system is
shown.

Name all the
parts indicated.

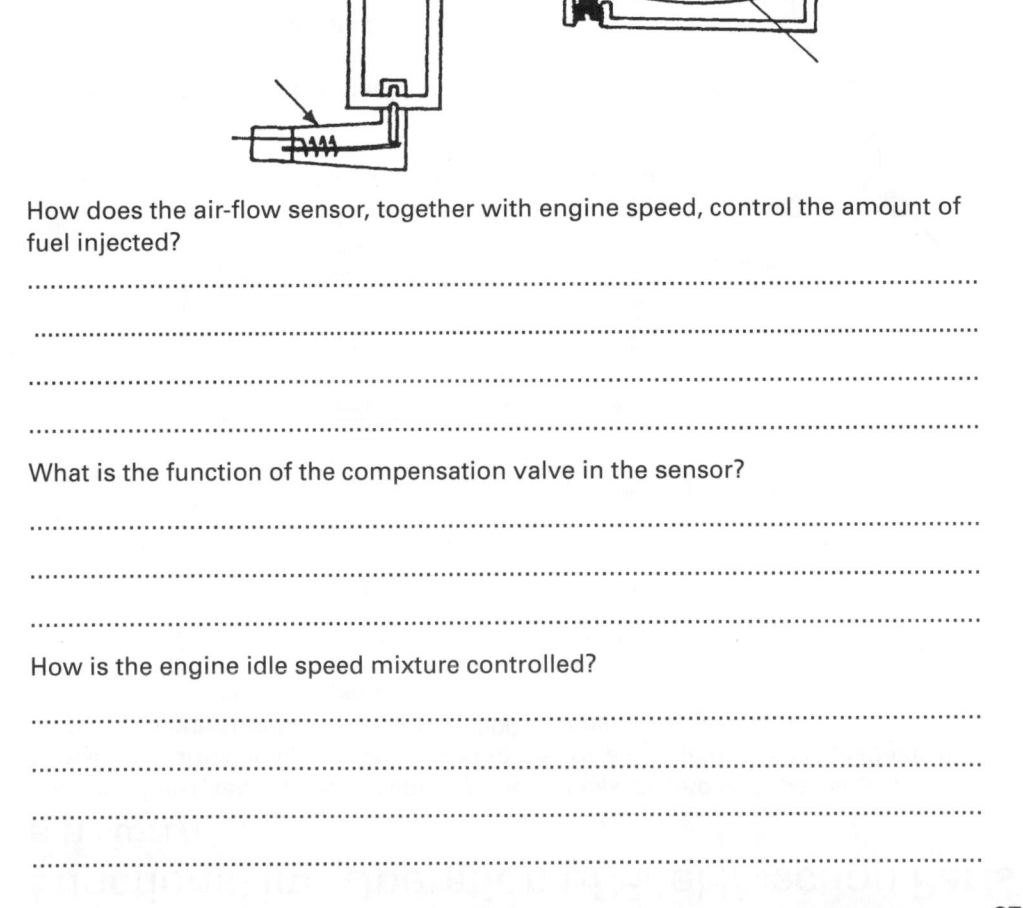

How does the air-flow sensor, together with engine speed, control the amount of fuel injected?

....

....

....

What is the function of the compensation valve in the sensor?

....

....

....

How is the engine idle speed mixture controlled?

....

....

....

....

The engine idle speed may be adjusted by the screw over the throttle valve.
The auxiliary air valve allows more metered air to enter to increase engine speed during the warm-up period.
The more modern types use a closed-loop idle speed control requiring no external idle speed adjustment.
Why is an air temperature sensor required? ....

....

....

....

(See also MASS FLOW SENSOR on page 240)

What controls Fuel Cut-off during no load cruising or engine braking?

....

....

BOSCH LE3 JETRONIC SYSTEM LAYOUT

What is the major difference between this LE3 Jetronic system and the L Jetronic?

....

....

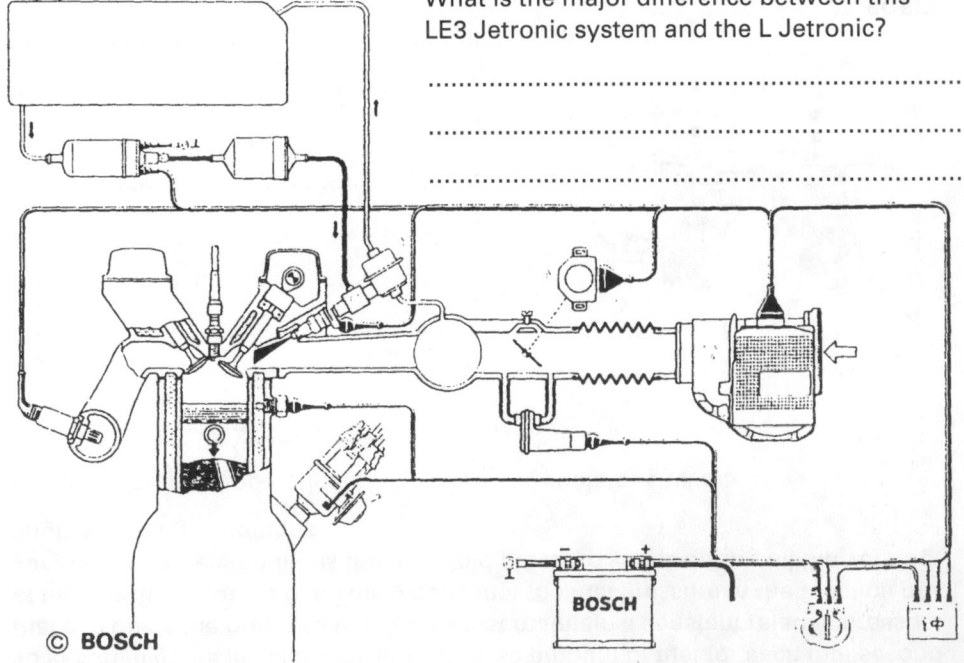

© BOSCH

270

Combined Ignition and Fuel Injection System

This is a more advanced type of Engine Management control. The ignition and fuel injection system are harmonised together through the control of a single microcomputer unit.

BOSCH MOTRONIC SYSTEM LAYOUT

Of the systems so far described the Motronic, while being electronically the most complicated, is the simplest to understand. The fuel injection is based on the Jetronic KE system.

How does the ignition obtain its operational information?

....

....

....

State the main advantages of this system as compared with other types.

1. ....   2. ....

3. ....   4. ....

5. ....   6. ....

State FOUR additional functions that may be included in the Motronic system.

1. ....

....

2. ....

....

3. ....

....

4. ....

....

Examine the Bosch Motronic schematic diagram shown below and name the numbered parts.

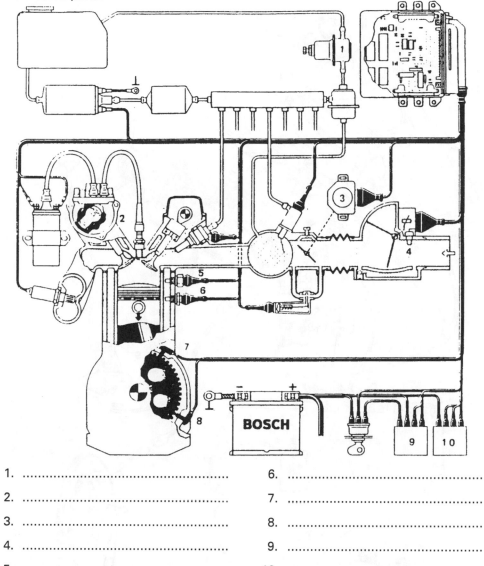

1. ....   6. ....

2. ....   7. ....

3. ....   8. ....

4. ....   9. ....

5. ....   10. ....

See page 240 for **ENGINE MANAGEMENT SENSORS**

MULTI-POINT FUEL INJECTION SYSTEM (ROVER)

The component parts shown are used on some Rover 820 series 2-litre twin overhead camshaft, sixteen-valve engines.

Recognise the components and sensors and name the parts numbered:

1. ....
2. ....
3. ....
4. ....
5. ....
6. ....
7. ....
8. ....
9. ....

10. ....
11. ....
12. ....
13. ....
14. ....
15. ....
16. ....
17. ....
   ....

What is the function of the inertia switch?

....

....

....

How does the idle-speed motor operate?

....

....

....

....

....

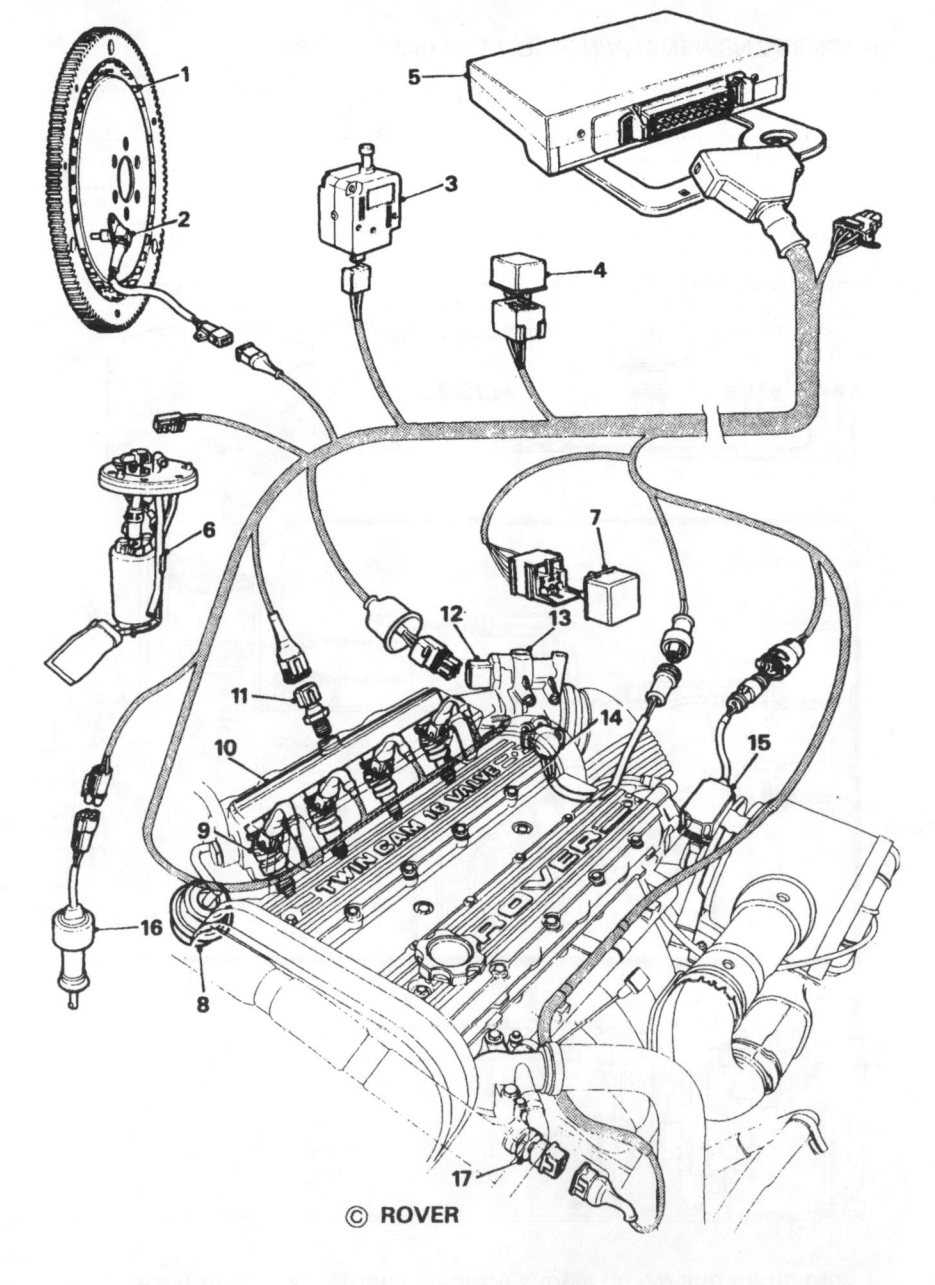

© **ROVER**

272

SINGLE-POINT FUEL INJECTION SYSTEM (ROVER)

The component parts shown are used on some Rover 820 series
2-litre twin overhead camshaft, sixteen-valve engines.

Recognise the components and sensors and name the parts
numbered:

1. ....

2. ....

3. ....

4. ....

5. ....

6. ....

7. ....

8. ....

9. ....

10. ....

11. ....

12. ....

13. ....

14. ....

15. ....

16. ....

17. ....

18. ....

19. ....

20. ....

21. ....

22. ....

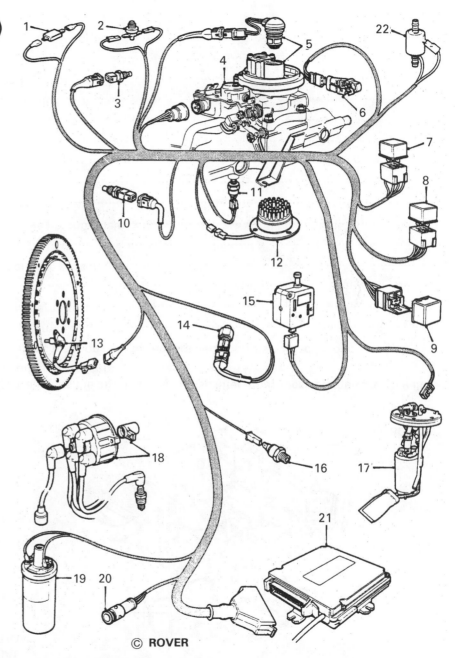

© ROVER

273

BOSCH MONO-JETRONIC SYSTEM LAYOUT

(THROTTLE BODY INJECTION)

In this system fuel is injected intermittently by a single solenoid-operated injector positioned above the throttle valve. The fuel distribution system via the inlet manifold is similar to that of the single carburettor layout.

Name the fuel system parts.

SCHEMATIC LAYOUT

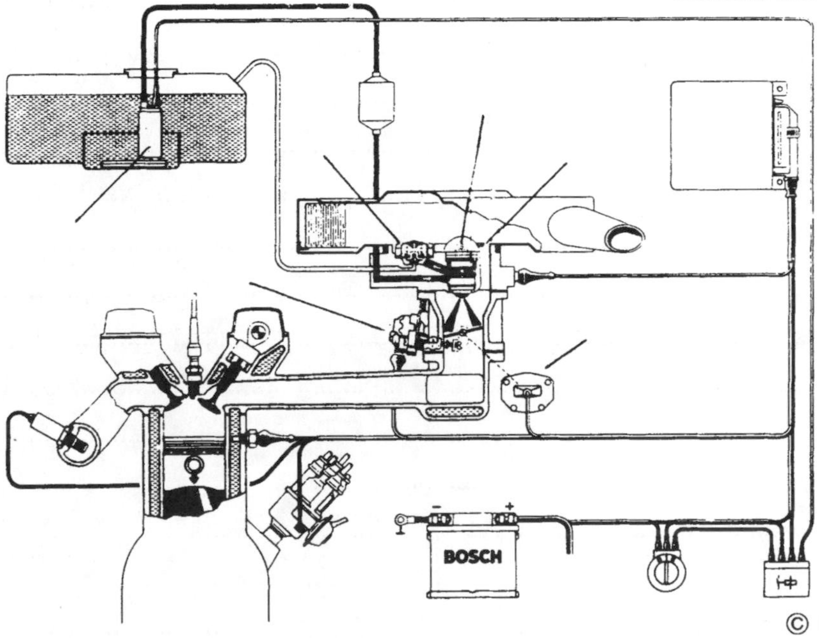

© BOSCH

Describe the main features of the fuel injection system:

The fuel pump ....

The primary pressure regulator ....

....

....

....

The injection unit ....

....

....

....

CENTRAL INJECTION UNIT

Name the main parts.

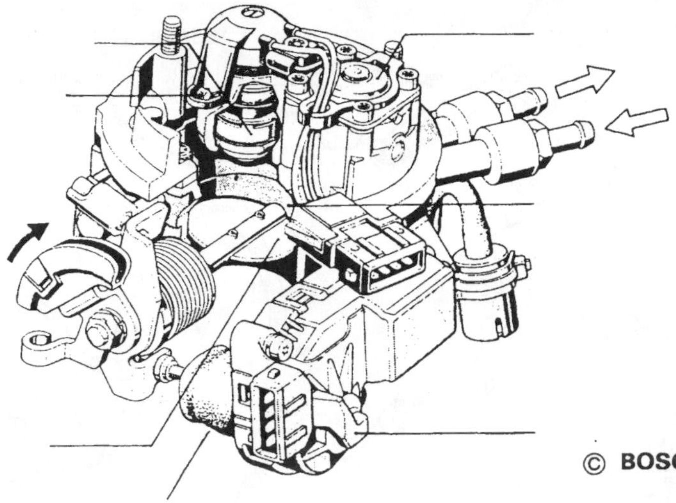

© BOSCH

The ECU transmits the injection time pulse after it has received the varying values from the different sensors.

State the basic functions of the sensors and controls named below:

Ignition system ....

Engine temperature sensor ....

....

Air temperature sensor ....

....

Throttle valve potentiometer ....

....

....

....

Cold start enrichment ....

Idle speed control ....

....

274

LAMBDA SENSOR

The Lambda closed-loop control using a Catalytic Converter is at present the most effective way of achieving low exhaust gas emission levels.

A schematic layout of the closed-loop control is shown below. Name the main parts and state the three products of combustion that the converter renders harmless.

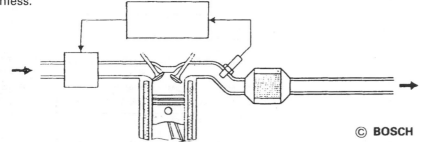

© BOSCH

The lambda sensor has been shown in position on many of the system layouts shown in this chapter. What is its basic function?

....

....

....

EXHAUST GAS RECIRCULATION (EGR)

Exhaust gas recirculation is a method commonly used to reduce the

....

from the exhaust emissions. It achieves this at part load running. The valve does not operate at idle and full load.

Show the direction of gas flow, name the basic parts and the sensors that inform the ECU when to operate the EGR valve.

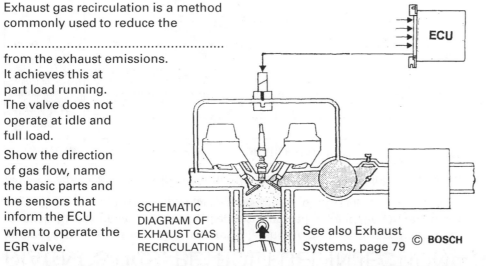

SCHEMATIC DIAGRAM OF EXHAUST GAS RECIRCULATION

See also Exhaust Systems, page 79   © BOSCH

ENGINE RUNNING PROBLEMS

It is essential to check all the related engine systems before assuming that a fuel injection system or electronic ignition system is the cause of an engine's bad starting or faulty running. (**Note**: The main items in the fuel injection and electronic ignition systems are the parts least likely to cause problems.)

List EIGHT preliminary checks that should be made before testing the engine management system:

1. ....

2. ....

3. ....

4. ....

5. ....

6. ....

7. ....

8. ....

List SIX preventive maintenance checks that should be observed. This could mean just checking and then if problems are found repairing and/or replacing.

1. ....

2. ....

3. ....

4. ....

5. ....

6. ....

When using voltmeters-ohm-meters, what care should be used when connecting them to the circuit?

....

....

....

....

Where sensors are concerned, what is the most likely problem?

....

DIAGNOSTICS: PETROL FUEL INJECTION – SYMPTOMS, FAULTS AND CAUSES

Engine Management servicing requires reference to manufacturers' workshop manuals to look up and follow the diagnostic procedure, and then to replace or repair the sensors, fuel pump, injectors, distributor or ECU etc.

Below is shown a fault finding chart for the BOSCH L & LE Fuel Injection System.

Repairs or adjustments should not be attempted without use of the proper manufacturer's equipment and instructions; such instructions are detailed and cover all diagnostic, removal and refitting procedures.

State probable causes for each symptom listed below.
Each cause will suggest any corrective action required.

SYMPTOM	PROBABLE CAUSES
Engine difficult to start or fails to start when cold	
Engine difficult to start or fails to start when hot	
Engine stalls or idles roughly when cold or warm	

SYMPTOM	PROBABLE CAUSES
Engine idles too fast	
Engine misses, hesitates, or stalls under load	
Engine low on power	
Engine failed emission test	

ENGINE SELF DIAGNOSIS

On modern engine management systems the ECU constantly monitors signals from the various sensors, compares them with values in the ECU memory, and detects incorrect readings when they occur as faults and stores them for reference. On the Audi 80/90 the faults can be indicated as flashing light codes on the engine fascia panel warning lamp. A fault code is made up of four flashing groups each with a maximum of 4 flashes and a 2.5 second pause between each group.

EIGHT sets of flash fault codes are indicated below.
Using the fault diagnosis table opposite state each code and the fault location.

Fault code ....

Location ....

Fault code ....

Location ....

Fault code ....

Location ....

Fault code ....

Location ....

Fault code ....

Location ....

Fault code ....

Location ....

Fault code ....

Location ....

Fault code ....

Location ....

Note: In practice only one or two faults would be expected at one time to require rectification.

FAULT DIAGNOSIS TABLE	Bosch KE/KE3-Jetronic		Audi 80/90
Fault Code	**Fault Location**	**Possible Cause**	**Remedy**
1-1-1-1	Control unit	Internal components	Replace control unit
2-1-2-1	Idle speed switch (Throttle switch I)	Faulty switch or shorted lead to terminal	Check switch
2-1-2-2	No engine speed signal from FEI unit	Broken lead between FEI terminal 17 & ECU terminal 30 or FEI unit faulty or Hall pick-up faulty	Check wiring Replace FEI unit Check Hall pick-up
2-1-2-3	Full load switch (Throttle switch II)	Switch stuck closed or lead shorted to terminal	Check switch & wiring
2-1-4-1	Knock control at knock limit	Engine pinking or knocking Fuel octane low incorrect ign. timing or knock sensor lead shield damaged	Check compression & fuel-injection system Use specified fuel Reset ign. timing Check sensor wiring
2-1-4-2	Knock sensor or knock detection	Break or short in sensor lead or Knock sensor defective or FEI not detecting pinking	Check knock sensor to FEI wiring Replace sensor Replace FEI unit
2-2-2-3	Altitude sensor	Break or short in wiring between sensor & control units or Defective sensor	Check wiring
2-2-3-2	Air-flow sensor potentiometer Load signal	Break or short in wiring between ECU & air-flow sensor or Break or short in wiring between ECU pin 21 & FEI pin 8	Check wiring & potentiometer
2-2-3-3	Reference voltage for load & altitude signals from ECU	Break in wiring between FEI pin 21 & ECU pin 26	Check wiring
2-3-1-2	Coolant temperature sensor	Break or short in sensor lead or Defective sensor	Check wiring Check sensor
4-4-3-1	Rotary idle adjuster valve	Break or short in lead between ECU & valve	Check wiring Check valve
4-4-4-4	No fault detected		
0-0-0-0	End of fault display sequence		*Autodata*

DIAGNOSING DIFFICULT FAULTS

All modern vehicles use Electronic Control Modules (ECM) – a renaming of the ECU – which, as well as controlling fuel and ignition output, may also control systems such as air conditioning, electronic gearbox, ABS braking and traction control. Any minor fault from any sensor or connection in these systems, since they are all interrelated, can cause very noticeable engine driveability problems which can be very difficult to trace. Fault code readers will only detect what they are programmed to recognise and for 10% of faults a multimeter or engine analyser is not up to the job, but an Oscilloscope, used by an experienced technician will most likely find the fault.

Hand Held Oscilloscopes

The modern portable hand held oscilloscope is a Digital Sampling Oscilloscope (DSO). What advantages does this type of oscilloscope have over the original analogue types which are commonly found in engine analysers (see page 206).

....
....
....

The waveform seen on the screen is a graphical representation, where the vertical axis is different scales of voltage (V), millivolts (mV), kilovolts (kV), and the horizontal scale is time in seconds (s), milliseconds (ms), microseconds (µs), nanoseconds (ns).
Like a voltmeter only two connections are needed for most tests. Also a trigger connection can be used, what is its purpose?

....
....
....
....

What function would an 'Autoset' allow?

....
....
....
....

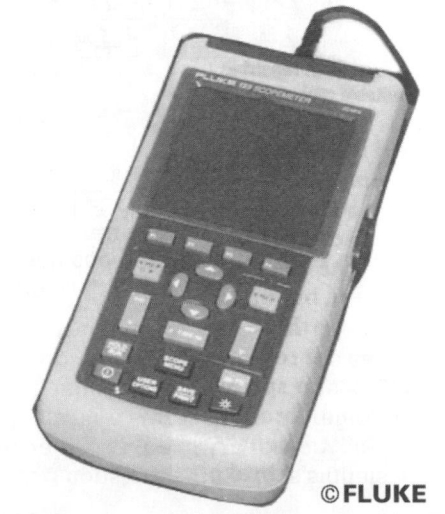
©FLUKE

Oscilloscope Patterns

Each electrical component has an identifiable oscilloscope trace and any deviation from this trace will most probably indicate a fault.
On the screens below draw in the correct expected tracers for the stated components. Comment on the shape of each waveform or trace.
All scale values are per division.

Scale 1V

Scale 10ms

Secondary ignition spark action

3V

5ms

Ignition-inductive pulse generator

....

2KV

5ms

Hall Effect pulse generator

10V

2ms

Injector

....

1V

100ms

Air flow meter

200mV

200ms

Lambda sensor

....

Scopes may have 2 or 4 channels. What system is a 4 channel scope ideal for checking? ....
....

278

When diagnosing faults you must understand the function of each part, how to isolate a problem and then correct it.

Into what FOUR basic categories do fuel injection problems fall when fault finding?

1. .... 2. ....

3. .... 4. ....

Pressure tests require running the fuel pump without running the engine (normally the electrical supply to the pump, if the engine is not running, is automatically switched off). How is such a fuel pressure test achieved?

....

....

....

....

When connecting a pressure gauge into the fuel system how should the residual pressure in the system, which could be over 3 bar, be released?

....

....

....

Describe how to test a component or sensor that is considered possibly faulty.

System make .... Vehicle ....

Test ....

1. ....

2. ....

3. ....

4. ....

5. ....

6. ....

7. ....

8. ....

....

SYSTEM PROTECTION DURING USE

Describe how the engine management fuel system should be protected during use or repair from the following hazards:

1. Ingress of dirt and moisture

....

....

....

....

2. Damage to sensors

....

....

....

3. Excessive heating of ECU

....

....

....

....

4. Damage by an external electrical supply

....

....

....

....

5 Removing electronic components

....

....

....

Chapter 13

Air Conditioning

Safety precautions	281	Manifold gauges	293
Basic refrigeration principles	282	Service ports	294
Basic refrigeration circuit	284	Refrigerant recovery	295
Major components	285	Evacuating a system	296
Fixed orifice tube system	288	Charging a system	296
Hoses	289	Temperature sensing – thermostatic switch	297
Oils	289	Leak detecting	298
Specialised components and electrical circuits	290	Performance testing	298
Pressure switches	290	Diagnostics	299
Evaporator fan circuit	291	Automatic climate control (ACC)	300
Condenser fan	292	Fascia control for a climate control system	300
Servicing tools/equipment	293	Air purifier system	300

SAFETY PRECAUTIONS

Air-conditioning systems are filled with refrigerants under high pressure. Since 1993 the refrigerant used in new vehicles is R134a, this replaced refrigerant R12 (freon). System fittings must not be loosened or components removed until the refrigerant has been correctly discharged, it is illegal to discharge refrigerant gases into the atmosphere. Refrigerants will absorb large amounts of heat given certain circumstances. Each type of refrigerant may have specific safety precautions to observe during its use. You should consult manufacturers for specific advice about their particular product. There are, however, a number of safety principles that apply to the use of all refrigerants.

Refrigerants can be dangerous in FOUR ways. Beside each listed item give a brief example of how the refrigerant is dangerous.

1. Skin contact

....

....

2. Contact with flame

....

....

....

3. Explosion

....

....

....

....

....

....

4. Suffocation

....

....

....

....

....

....

List the personal protective clothing that should be worn when working on air-conditioning systems.

....

....

Refrigerants are stored in pressurised containers like the one shown in the diagram. It is important that these containers are treated carefully. Never allow the temperature of the containers to rise above 50°C.

Refrigerant container

The containers are colour-coded, to indicate the type of refrigerant stored. List the colour code for containers of each refrigerant below.

R12 ....

R134a ....

Why is it very important that different types of refrigerant are not mixed?

....

....

SAFETY PRECAUTIONS

Different refrigeration servicing equipment and parts should be used for each type of refrigerant. It is therefore necessary to have separate sets of equipment for each type of gas serviced in the workshop.

On modern vehicles using R134a different fittings requiring different fitting tools are used so there should be no chance of parts becoming mixed.

What other precautions should be taken to ensure that different refrigerants are not mixed?

....

....

....

R134a is a HydroFluoroCarbon (HFC) gas.

R12 is a ChloroFluoroCarbon (CFC) gas.

Explain why refrigerant should be collected and recycled.

....

....

....

....

....

....

The air conditioning system must be professionally discharged using the correct equipment before any part of the system is replaced.

State for safety reasons TWO other conditions when the system should be discharged while the vehicle is being repaired.

1. ....

....

2. ....

....

BASIC REFRIGERATION PRINCIPLES

To understand air-conditioning principles you must understand principles relating to heat and matter.

Matter is defined as anything that occupies space and has weight. All things are composed of matter and are found in one of three forms. State the three forms of matter.

1. ....   2. ....   3. ....

If heat is added to or removed from matter it changes state. For example, if heat is removed from water it turns to .... and if sufficient heat is added it changes to ....

HEAT

Heat is a form of energy that can be transferred from one place to another. The transfer of heat cannot take place unless there is a difference in temperature between the two objects. Heat energy travels in one direction only, from a warmer object to a cooler object.

The application or removal of heat does not always result in a change of temperature, although other changes may occur. Particular names are used to describe the different effects of heat. Define the following names of heat.

Sensible heat is ....

....

....

Latent heat is ....

....

....

The latent heat of vaporisation is ....

....

....

The latent heat of condensation is ....

....

REFRIGERANT CIRCUIT OF AN AIR-CONDITIONING SYSTEM

Name the parts indicated on the circuit.

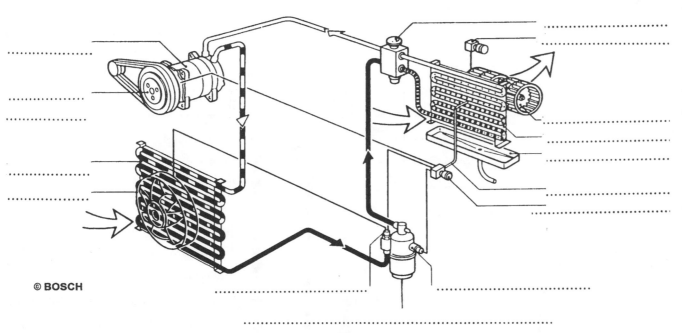

© BOSCH

State the purpose of a vehicle's air condition system.

....

....

....

....

....

HEAT LOAD

Heat load is the amount of heat the air conditioning system must overcome to cool the vehicle. Heat load for vehicles can vary depending on things such as colour of vehicle and the amount of glass area.

List other factors that will affect vehicle heat load.

....

....

BASIC OPERATION OF THE AIR-CONDITIONING SYSTEM RECEIVER/DRIER TYPE

The air conditioning system operates on a vapour, compression, refrigeration cycle.

The compressor, which is belt driven from the engine's crankshaft pulley, is operated by a magnetic clutch. The compressor raises the pressure and therefore the temperature of the refrigerant. High pressure vapour leaves the compressor and travels to the condenser.

The condenser is fitted at the front of the vehicle and is cooled, like a normal radiator, by air flow generated by the fan blowing over the fins and by ram air caused by vehicle movement. The vapour passes through long thin tubes in the condenser and is cooled and condensed into a high pressure liquid. This refrigerant liquid then travels to the receiver/drier where it is temporarily stored in a fluid reservoir, filtered, and any moisture in the liquid is absorbed by the desiccant or silica gel crystals.

The refrigerant continues to the expansion valve where it passes through a small orifice which restricts its flow. This in combination with 'suction' from the compressor results in a reduction in the pressure of the refrigerant in the evaporator.

The orifice as it sprays into the tubes of the evaporator atomises the refrigerant and at the same time the relatively warm air in the vehicle, which is blown over the evaporator fins, causes the refrigerant to boil. As it changes state from a low pressure liquid to a low pressure vapour it absorbs heat and enables the air in the vehicle to be cooled.

The refrigerant continues around the system and is again drawn into the compressor.

BASIC REFRIGERATION CIRCUIT

Compressor ....
....
....
....

Evaporator ....
....
....
....
....
....
....
....
....
....

TX Valve ....
....
....
....
....
....
....

Show the direction of refrigerant flow and indicate where the changes of state take place.

Complete the text boxes by explaining what is occurring to the refrigerant in each component.

Compressor

Condenser ....
....
....
....
....

Evaporator

Receiver / Drier ....
....
....
....
....
....

Condenser

TX Valve

Receiver/ Drier

KEY

High pressure vapour
High pressure liquid
Low pressure vapour
Low pressure liquid

MAJOR COMPONENTS

Compressor

The compressor, sometimes called the heat pump, circulates the refrigerant and at the same time raises the pressure and therefore the temperature of the gas. There are THREE basic types of compressors that can be used in automotive air-conditioning systems. A reciprocating piston type which may have from 1 to 10 pistons, these are used for heavy duty work. A rotary vane type, where, as the compressor shaft rotates, the vanes and housing form chambers which individually reduce in size to compress the gas.

Describe a third type of compressor.

....

....

....

....

Identify the parts indicated on the compressor shown.

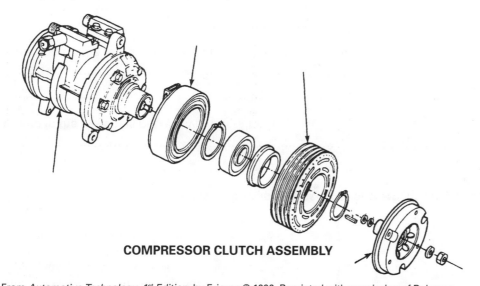

COMPRESSOR CLUTCH ASSEMBLY

From *Automotive Technology*, *1ˢᵗ Edition*, by Erjavec © 1992. Reprinted with permission of Delmar a division of Thomson Learning.

Why is an electromagnetic clutch fitted to the compressor?

....

....

....

....

....

....

Condenser

The condenser acts just like a conventional radiator and the refrigerant flowing through the fine tube in the condenser is cooled by fan-forced air and ram-air when the vehicle is moving.

Indicate the direction of refrigerant flow and name the parts shown.

From *Automotive Technology*, *1ˢᵗ Edition*, by Erjavec © 1992. Reprinted with permission of Delmar a division of Thomson Learning.

Where on the vehicle is the condenser fitted?

....

....

....

Explain what happens to the refrigerant as it passes through the condenser.

....

....

....

Receiver/drier

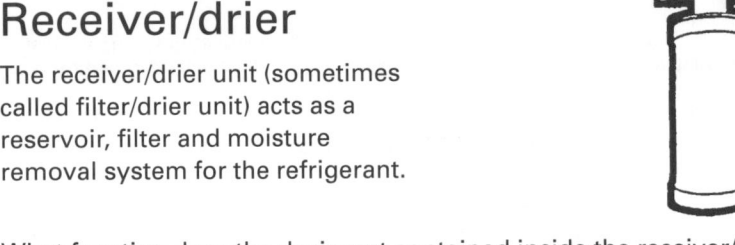

The receiver/drier unit (sometimes called filter/drier unit) acts as a reservoir, filter and moisture removal system for the refrigerant.

What function does the desiccant contained inside the receiver/drier perform?

....
....

Where in the system is the receiver/drier fitted?

....
....

What state is the refrigerant in as it passes through the receiver/drier?

....

Look at the diagram below. It shows a cross-sectional view of a receiver/drier. Label each of the items that has an arrow pointing to it. Also identify the hose connections, e.g. in/out, and use arrows to identify the flow of refrigerant through the receiver/drier.

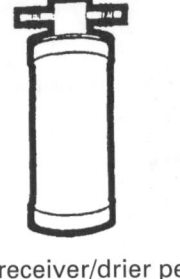

When only one type of refrigerant was used (R12), there was only one type of desiccant used (XH-5). The advent of new types of refrigerant has required different types of desiccant to be used. Complete the table below to show which type of desiccant can be used with each type of refrigerant.

Type	R12	R134a	Various Blends
XH-5	....	....
XH-7	....	....
XH-9	....	....

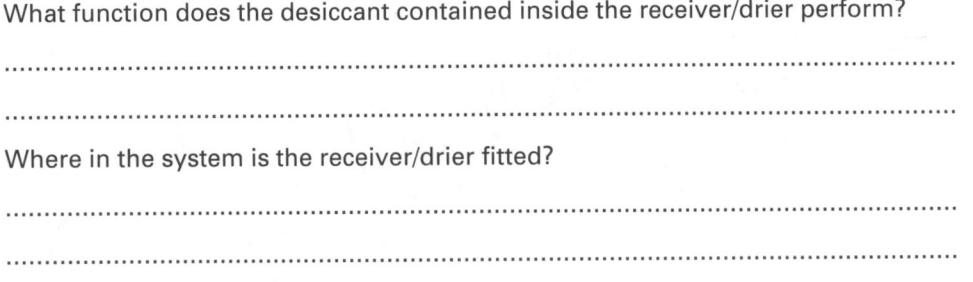

From *Automotive Technology*, *1ˢᵗ Edition*, by Erjavec © 1992.
Reprinted with permission of Delmar a division of Thomson Learning.

Sight Glass

The sight glass allows you to see the flow of refrigerant. It can be located on the receiver/drier, as shown above, or in-line, shown opposite. The diagram opposite shows the appearance of the refrigerant under certain conditions. Complete the table below to identify these conditions.

Condition	Indication
Clear	....
Foam or Bubbles	....
Streaky	....
Cloudy	....

TX Valve (Expansion Valve)

The TX (Thermal Expansion) valve has a small orifice though which the refrigerant must pass. This orifice presents a flow restriction and so regulates the amount of refrigerant that passes through to the evaporator.

Describe what happens to the refrigerant as it passes through the TX valve.

....

....

What also aids pressure reduction to the evaporator?

....

....

Look at the diagram below. It shows the cross-sectional view of a TX valve. Label each of the items indicated.

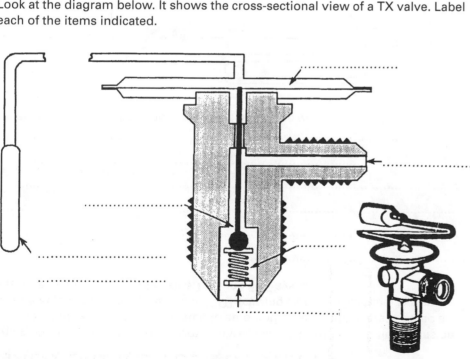

The temperature sensing bulb and capillary tube are integral to the operation of the TX valve shown above. Describe the operation of the sensing bulb and capillary tube.

....

....

....

....

....

....

....

....

Evaporator

The evaporator has a winding tube construction surrounded with fins, it is designed to absorb the maximum amount of heat in the smallest possible space. The warm air flows through the fins and over the tubes containing the refrigerant.

Describe what happens to the refrigerant as it passes through the evaporator.

....

....

Explain how the evaporator dehumidifies the cabin air.

....

....

....

....

....

....

FIXED ORIFICE TUBE SYSTEM

The system described so far is commonly called a TXV (thermostatic expansion valve) system. Another system fitted to automotive air conditioners is called a FOT (fixed orifice tube) system or a CCOT (cycling clutch orifice tube) system. Describe the difference between a TXV and FOT system.

....

....

....

....

....

Label the components on the basic FOT system shown,

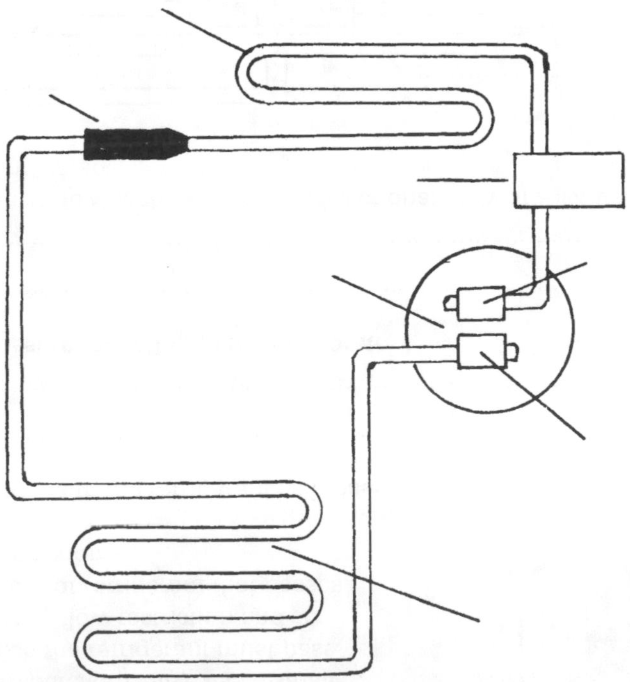

Note. A refrigerant sight glass is not used with this system.

The diagram shows a fixed orifice tube system. The orifice tube or venturi is positioned in the inlet to the evaporator.

The accumulator/dehydrator is positioned just before the compressor. What is its function?

....

....

....

Note the quick release Schrader valve couplings 1 high pressure side and 2 low pressure side. Why are these valves positioned high up on the pipe lines?

....

....

What is the joint function of the pressure regulating switch 3 on the high pressure side and the pressure cycling switch 4 on the low pressure side?

....

....

....

....

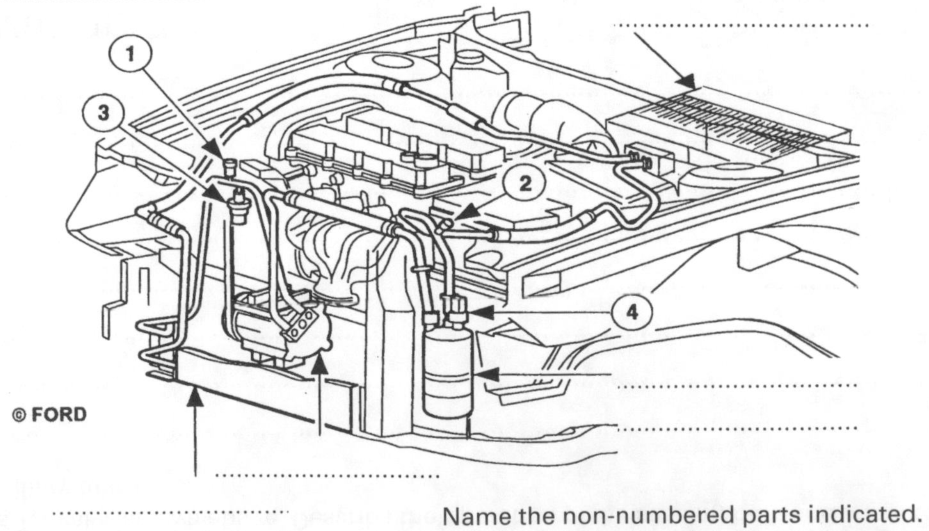

....

© FORD

Name the non-numbered parts indicated.

HOSES

Hoses are an integral part of the air-conditioning system. With new types of refrigerant it is extremely important to fit the correct type of hoses.

State the type of hose that should be fitted to automotive air-conditioning systems.

....

....

What would be the consequences of not fitting the correct types of hose to systems gassed with R134a?

....

R12 gas has not been available since the year 2000. When repairs are needed, vehicles using R12 will have to be converted to using R134a. This will involve changing the receiver/drier, all rubber hoses and O-rings, the compressor oil, then having the system flushed out and recharged.

When checking older vehicles, systems with snap-type service ports are already R134a compatible, but if they have threaded ports it means the system uses or originally used R12.

OILS

Various refrigeration components require lubrication. For example the compressor has many friction surfaces such as bearings and pistons that must be lubricated. The oil in a refrigeration system is moved around the system by the refrigerant.

What damage would result to an air-conditioning system if the compressor was still operated after all the refrigerant had leaked from the system?

....

....

....

....

To prevent the compressor operating when there is no refrigerant in the system a low pressure cut-out switch can be used. Describe how this switch would prevent damage to the air-conditioning system.

....

....

....

....

Most R12 air-conditioning systems use a 5GS mineral oil. This was a common air-conditioning oil before the advent of the new types of refrigerant. New types of refrigerant require different types of oil. The different types of refrigerant oil do not mix very well. As a consequence you must be extremely careful not to mix incompatible oils when working on systems.

What would happen if incompatible refrigerant oils are mixed in a system?

....

....

....

....

List three different types of refrigerant oil currently used in automotive air-conditioning systems.

1. .... 2. .... 3. ....

No matter what type of refrigerant oil is used they are generally hygroscopic. That is, they absorb moisture. Moisture in an air-conditioning system will impair its performance and can cause component damage. What precautions should you take to ensure that moisture is not absorbed by oil?

....

....

....

SPECIALISED COMPONENTS AND ELECTRICAL CIRCUITS

The refrigeration circuit discussed so far has only been very basic. Many additional components and an electrical system are also fitted to enhance the operation of the system. The basic electrical circuit for an air conditioner can be broken into three main circuits: they are the clutch, fan or blower and condenser fan circuits. As you progress through this section you will gradually build more complex electrical circuits until you have a complete circuit for a basic system. First starting with the clutch circuit. The clutch is used to stop the compressor from pumping continuously.

Complete the simple clutch circuit below.

Thermostatic switch

Clutch coil

Fuse

Air-con. switch

+ −

12V

Describe what would happen to the evaporator if a clutch was not fitted to the compressor.

....

....

....

....

PRESSURE SWITCHES

Low and high pressure switches are often used on air-conditioner systems to enhance the system operation. Each performs a specific role and is connected in series with the clutch circuit. In the space provided explain the reason for fitting low and high pressure switches.

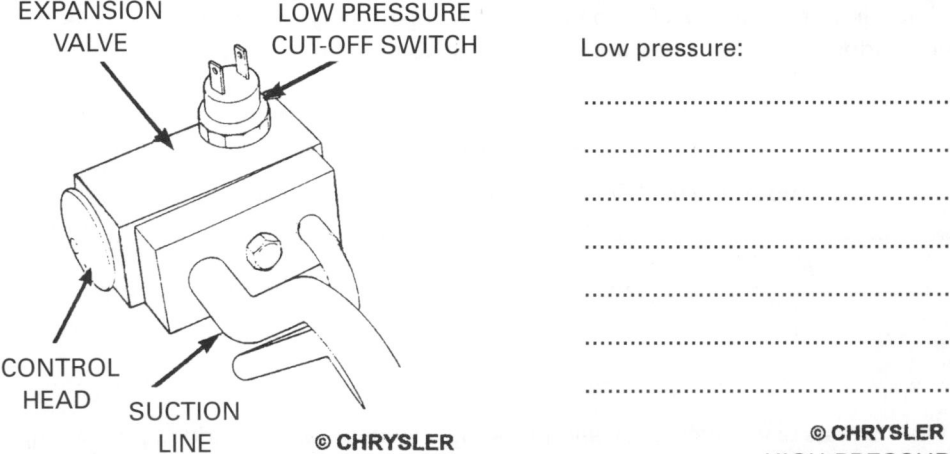

EXPANSION VALVE

LOW PRESSURE CUT-OFF SWITCH

CONTROL HEAD

SUCTION LINE

© CHRYSLER

LOW PRESSURE SWITCH FITTED ON EXPANSION VALVE

Low pressure:

....

....

....

....

....

....

High pressure:

....

....

....

....

....

© CHRYSLER
HIGH-PRESSURE CUT-OUT SWITCH

HIGH-PRESSURE RELIEF VALVE

COMPRESSOR MANIFOLD

HIGH PRESSURE SWITCH FITTED ON COMPRESSOR

The circuit diagram below includes a high and low pressure switch in the circuit. Complete the diagram.

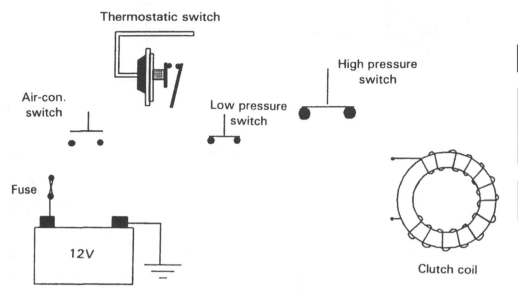

Sometimes instead of a high and low pressure switch a binary switch is fitted. Explain what a binary switch is.

....

....

Complete the diagnostic flow chart for if the compressor would NOT engage.

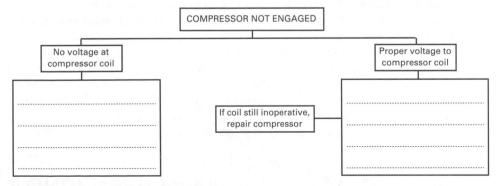

Complete the diagnostic flow chart for if the compressor engages but does NOT operate.

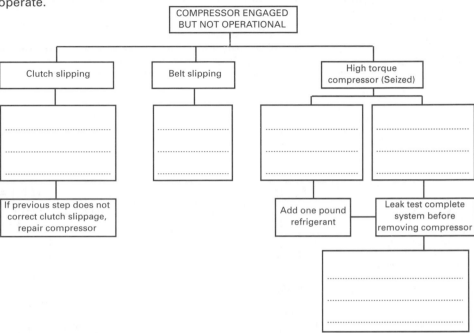

EVAPORATOR FAN CIRCUIT

The evaporator fan forces cabin air through the evaporator thus ensuring the air cools. Evaporator fans are usually multispeed. They are often three-speed but sometimes four or variable speed fans. The multiple speeds are obtained by inserting resistances into the circuit. Complete the fan circuit below.

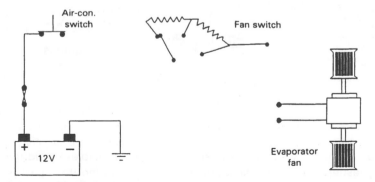

CONDENSER FAN

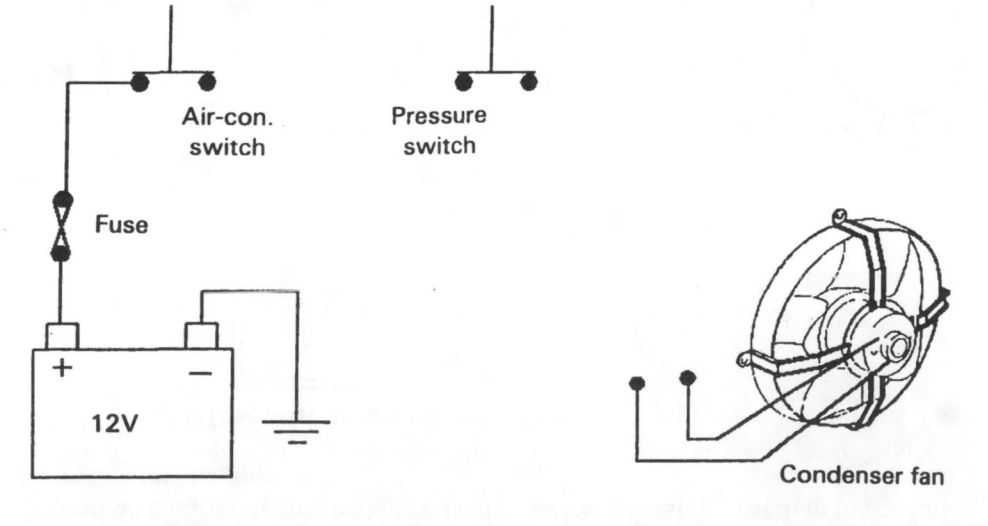

Electric condenser fans are used to increase the amount of ram air flowing through the condenser. Describe what would happen to the air-conditioning system if not enough ram air is flowing through the condenser.

....

....

....

Different manufacturers have different methods for controlling when electric condenser fans switch on. For example some switch on when the air-conditioning main switch is operated, others have a pressure switch that turns the fan on if the pressure of the refrigerant rises too high. The circuit below uses a pressure switch to turn the fan on and off. Complete the circuit.

Quite often relays are also used in the air-conditioning electrical circuit. They are used to reduce the amount of voltage drop in the clutch and condenser fan circuits. Complete the diagram so that relays are used to switch the clutch and fan circuits.

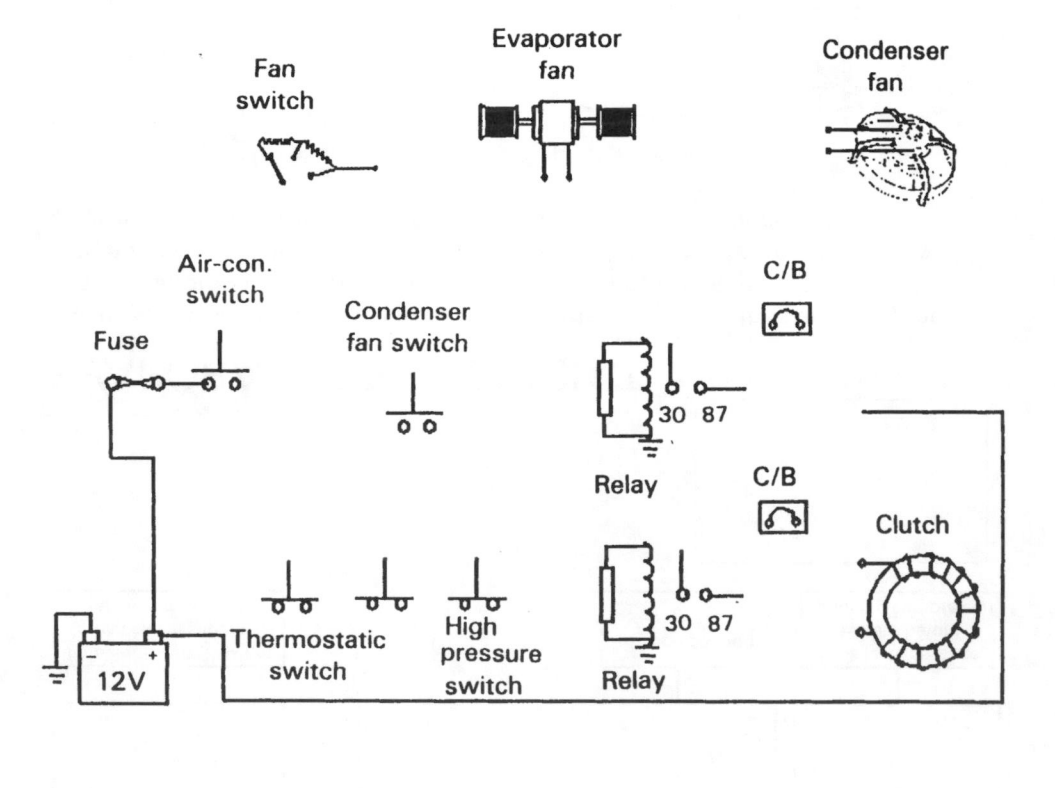

The relays used on the diagram are spike protected. Explain why this type of relay is used on modern motor vehicles.

....

....

....

....

SERVICING TOOLS/EQUIPMENT

Some of the servicing tools used for automotive air-conditioning systems are manifold gauges, vacuum pumps, reclaiming systems, recycling/reclaimers, lead detectors and service valves.

Each type of refrigerant will require a different set of servicing equipment. Explain why.

....

....

....

....

Many refrigerant bottles will have two access valves fitted to the top. They will usually be colour-coded and marked as to their use. For example one will be vapour and other liquid refrigerant. It is important that the hoses are connected to the valve appropriate to the state of refrigerant required i.e. vapour or liquid.

Which colour will be associated with:

Vapour ....

Liquid..

If the bottle you are using only has a single valve how would you remove liquid refrigerant from the bottle?

....

....

MANIFOLD GAUGES

Manifold gauges are used to diagnose faults and control the flow of refrigerant when gassing and degassing air-conditioning systems. It is very important that the manifold gauges are correctly connected to an air-conditioning system.

Complete the diagram to show how the manifold gauges should be connected to an air-conditioning system.

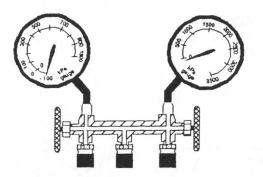

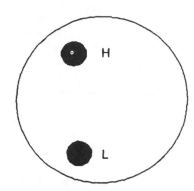

You should never open the high side valve on the manifold gauges while the air-conditioner is operating. Explain why.

....

....

....

Non-return valves should be used on the hoses of the manifold gauge. They are fitted on the end of the hose that connects to the air-conditioning system. Explain why non-return valves are used.

....

....

....

....

....

SERVICE PORTS

A number of different service ports are found on automotive air conditioners, they provide access to the refrigerant system to recover, recycle or replace the refrigerant and to allow testing of the system.

Most service ports have a schrader valve fitted which is similar in operation to tyre valves. R12 and R134a valves have different fittings. This is to ensure that refrigerants cannot be accidentally mixed. Look at the diagram below. Two different types of fittings are shown. Identify the type of refrigerant each is used on.

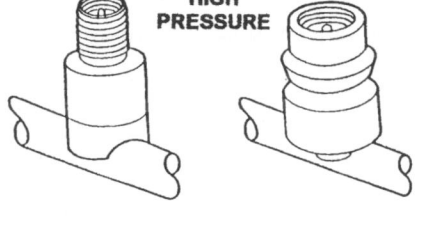

There are generally two service ports, one gives access to the high pressure side while the other allows access to the low pressure side of the system. Service ports can be fitted in different positions in the system. In older systems they were commonly found at the rear of the compressor. On modern systems the compressor can be difficult to gain access to, therefore the ports are often positioned on other locations.

Describe other places that service ports can be fitted.

....

....

....

....

Identify the high side and low side connections on both drawings below.

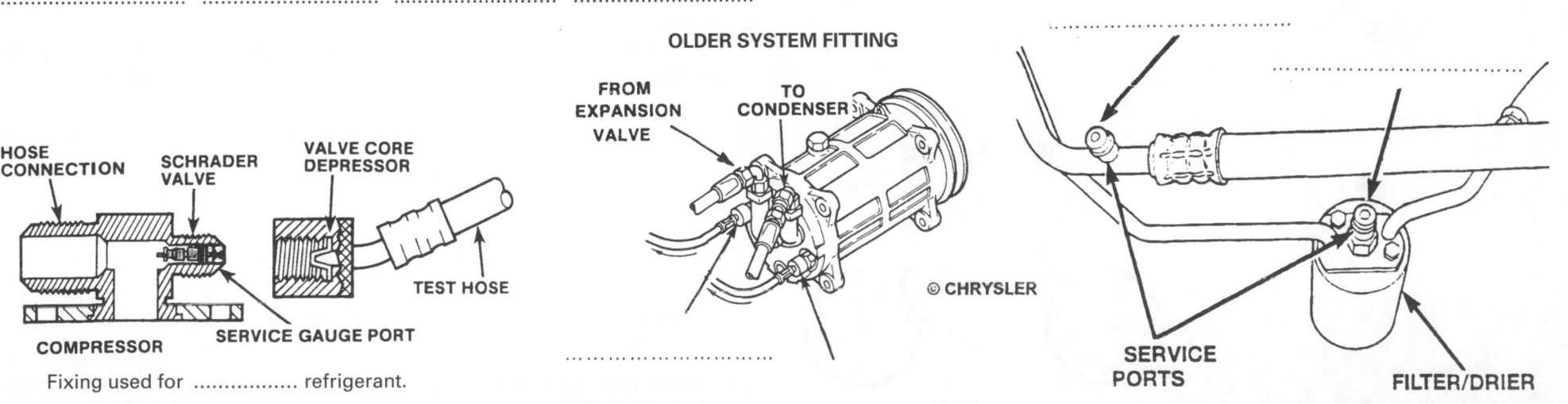

Fixing used for .... refrigerant.

From *Automotive Technology, 1st Edition*, by Erjavec © 1992. Reprinted with permission of Delmar a division of Thomson Learning.

See also page 288 for position of service ports.

REFRIGERANT RECOVERY

There are many different manufacturers of recovery/recycle units; each will have specific instructions relating to their use. Consult manufacturers' instructions before attempting to use any equipment you are unfamiliar with. Two basic types of recovery units are:

(a) recovery and recycle units
(b) recovery only units.

Explain the difference between the two types of units.

....

....

....

....

....

....

....

R-134 REFRIGERANT RECOVERY MACHINE

© CHRYSLER

REFRIGERANT RECOVERY/RECYCLING STATION

The diagram below shows a typical recovery unit and a system compressor with service valves. Complete the diagram by hooking up the hoses to the recovery unit, manifold gauges and the compressor. Indicate the colour of the hoses.

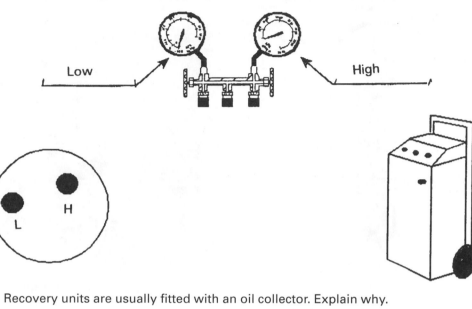

Low   High

Recovery units are usually fitted with an oil collector. Explain why.

....

....

....

ADDING OIL

Some system oil is lost when system components are replaced and refrigerant is reclaimed. This oil must be replaced. Manufacturers' specifications should be consulted for correct amounts of oil. List approximate amounts of oil that should be added if the following components are replaced.

Condenser: ....

Receiver/Drier: ....

Evaporator: ....

EVACUATING A SYSTEM

Explain why systems need to be evacuated before charging, and what occurs during the process.

....

....

....

....

....

....

State a typical amount of time for which the system should be evacuated.

....

....

....

Connections for evacuation equipment will depend on the type of equipment used in your workshop. An example is shown below. Complete the hose connections on the diagram. Indicate the position the hand valves would need to be in, i.e. open/closed, in order to evacuate the system. Also, using arrows indicate the direction of gas flow during the process.

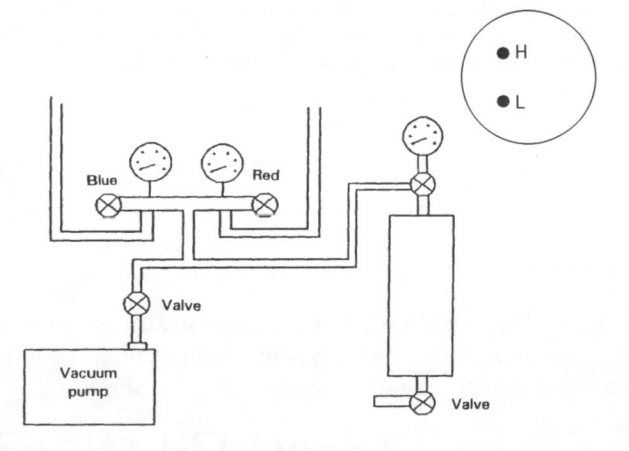

CHARGING A SYSTEM

When charging an air-conditioning system the amount of refrigerant placed in the system needs to be measured. In an R12 system this would often be done by monitoring the sight glass of the receiver/drier. However with other types of refrigerants and modern systems this procedure is not accurate enough. The preferred way of measuring the amount of gas going into a system is by using a graduated measuring cylinder or scale.

Explain how a graduated measuring cylinder works.

....

....

....

....

....

....

Identify the components shown below.

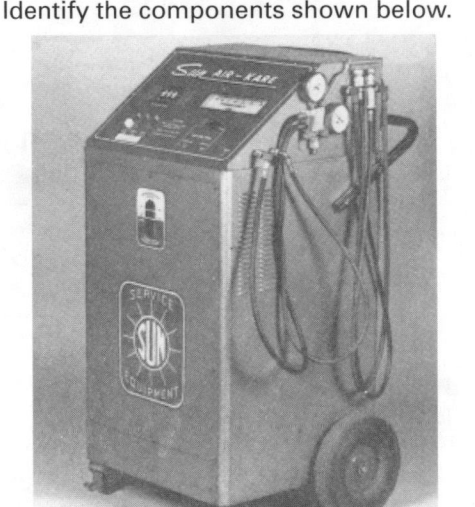

....

From *Automotive Technology, 1st Edition*, by Erjavec © 1992. Reprinted with permission of Delmar a division of Thomson Learning.

Describe how a measuring weigh scale can be used to check the weight of the refrigerant as it is passed into the system.

....

....

....

....

....

Caution must be observed when liquid filling. If liquid refrigerant enters the low side of the compressor it can cause serious damage to compressor components. To ensure no damage occurs, use the hand valve to regulate the amount of refrigerant entering the compressor so that low side pressure does not exceed 270 kPa (40 psi).

Complete the diagram below to show the hose connections to liquid fill an air-conditioning system using a graduated cylinder.

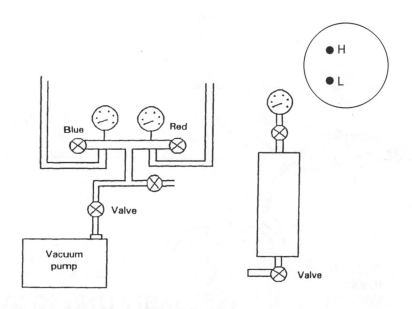

Explain how you would check for blockages in the system as you charge.

....

....

....

....

....

TEMPERATURE SENSING – THERMOSTATIC SWITCH

Adjustment to the thermostatic switch may be necessary to ensure correct operation of the system, especially if a new switch is fitted.

Describe how a thermostatic switch is adjusted.

....

....

....

....

....

....

....

....

....

....

....

....

....

Check a manufacturer's procedure and find out what temperature the thermostatic switch should:

cut in at ....

cut out at ....

TEMPERATURE SENSING PROBE

LEAK DETECTING

Leak detecting is a very important service procedure. Two basic types of leak detector can be used on modern systems. They are electronic detectors and dye type detectors.

Describe how each is used.

Electronic

....
....
....
....
....
....

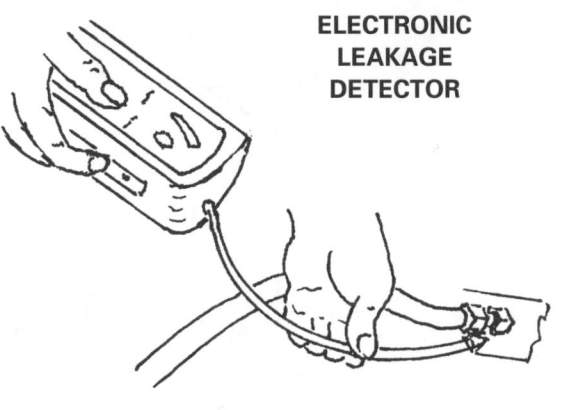

ELECTRONIC LEAKAGE DETECTOR

DYE INJECTION TOOL

ULTRA-VIOLET LAMP

Dye

....
....
....
....
....
....
....

PERFORMANCE TESTING

Manufacturer's specifications should always be consulted when performance testing a vehicle. Describe a typical performance testing procedure.

....
....
....
....
....
....
....
....
....
....
....
....
....
....
....

DIAGNOSTICS: AIR CONDITIONING

Diagnosis of air-conditioning systems requires that you read the manifold gauges correctly and look for other indications as to the performance of the system. This will require considerable practical experience. Complete the partially completed diagnostic table.

COMPLAINT	LOW PRESSURE GAUGE READING	HIGH PRESSURE GAUGE READING	OTHER SYMPTOMS	DIAGNOSIS AND CORRECTIVE ACTION
Little or no cooling	low	low	Bubbles in sight glass evaporator, air cool but not cold	Low refrigerant, possible leak. Leak detect and repair faults before regassing.
Little or no cooling	high	high	Occasional bubbles in sight glass, evaporator cool but not cold	....
Little or no cooling	very low into vacuum	low	TX valve is cold or frosted, air discharge is not cold	....
Insufficient cooling or no cooling	high	high	Evaporator sweating or frosted, suction hose sweating or frosted	....
Little or no cooling	....	....	....	Condenser clogged with debris. Clean condenser.
Little or no cooling	....	....	....	Compressor reed valves faulty. Replace compressor reed valves.
Degreased cooling efficiency	....	....	....	Overcharge of refrigerant. Remove some refrigerant.
Little or no cooling after 5–15 minutes	....	....	....	Excessive moisture in system. Reclaim refrigerant replace receiver/drier and evacuate before regassing.

AUTOMATIC CLIMATE CONTROL (ACC)

Recent developments in computer technology have seen manufacturers develop air-conditioning systems that once set automatically control the cabin environment. This system basically has an ECU that receives signals from a number of inputs from different sensors and provides outputs to control cabin temperature and humidity regardless of variables such as ambient air temperature.

Complete the diagram by naming inputs and outputs for a climate control system.

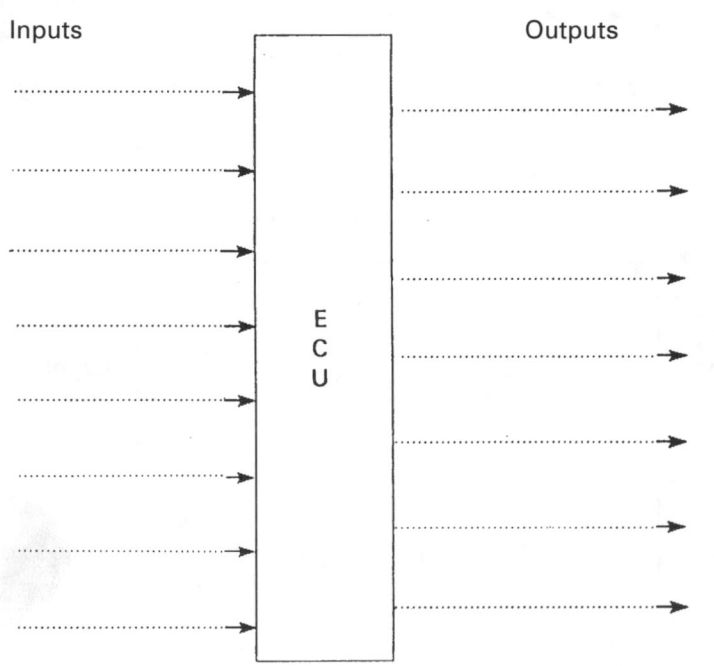

Inputs   Outputs

E C U

On damp days a common driving problem is the misting up of the windows. The cold glass condenses the moisture in the air in the vehicle and a dangerous loss of visibility can occur. How does a Climate Control System overcome this problem?

....

....

....

FASCIA CONTROL FOR A CLIMATE CONTROL SYSTEM

Diagram shows the driver's control for a fully automatic air conditioning system. All control is by electrical switches and the target temperature is displayed in digital form. Identify the switches indicated.

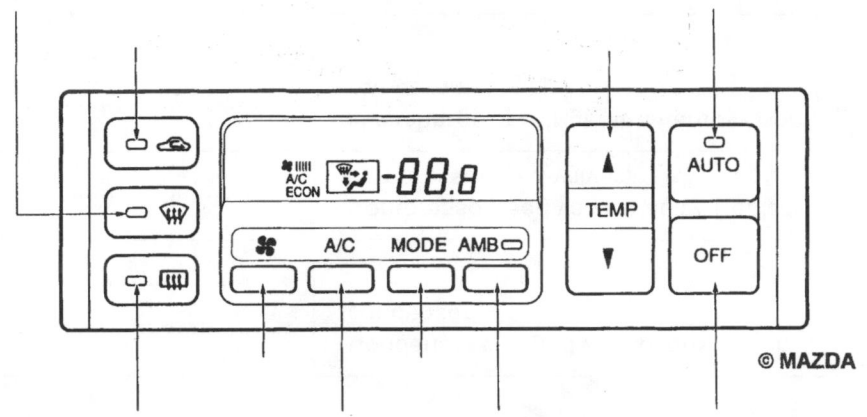

© MAZDA

AIR PURIFIER SYSTEM

The schematic drawing below show how traffic pollutants can be removed from the air before it enters the vehicles interior.

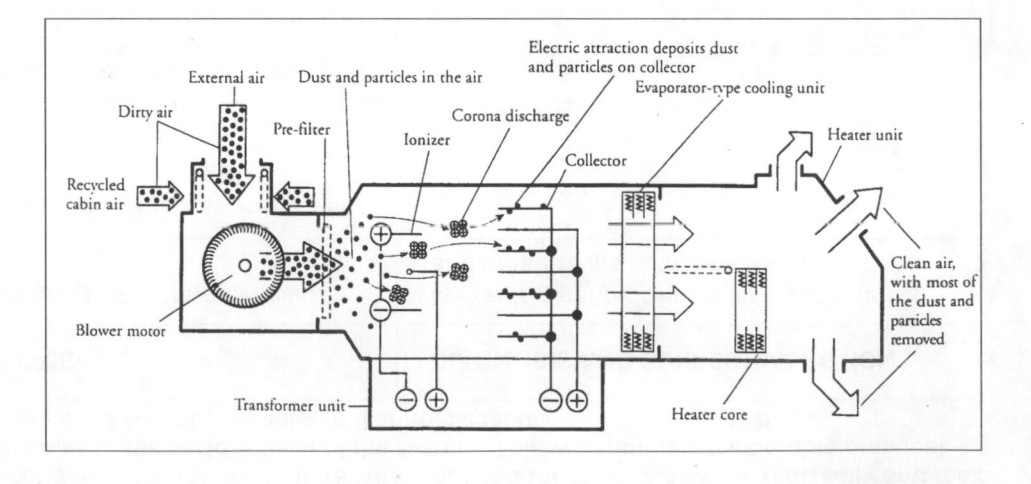